High Temperature Polymer Blends

Related titles:

Advances in shape memory polymers
(ISBN 978-0-85709-852-8)

Advances in filament yarn spinning of textiles and polymers
(ISBN 978-0-85709-499-5)

Smart polymers and their applications
(ISBN 978-0-85709-695-1)

High Temperature Polymer Blends

Edited by
Mark T. DeMeuse

AMSTERDAM • BOSTON • CAMBRIDGE • HEIDELBERG • LONDON
NEW YORK • OXFORD • PARIS • SAN DIEGO
SAN FRANCISCO • SINGAPORE • SYDNEY • TOKYO
Woodhead Publishing is an imprint of Elsevier

Woodhead Publishing is an imprint of Elsevier
80 High Street, Sawston, Cambridge, CB22 3HJ, UK
225 Wyman Street, Waltham, MA 02451, USA
Langford Lane, Kidlington, OX5 1GB, UK

British Library Cataloguing-in-Publication Data
A catalogue record for this book is available from the British Library.

Library of Congress Control Number: 2013957504

ISBN 978-1-84569-785-3 (print)
ISBN 978-0-85709-901-3 (online)

For information on all Woodhead Publishing publications visit our website at http://store.elsevier.com/

Typeset by Newgen Knowledge Works Pvt Ltd, India

Contents

Contributor contact details *xi*

1 **Introduction to high temperature polymer blends** 1
M. T. DeMeuse, MTD Polymer Consulting, USA

1.1 Introduction 1
1.2 General principles of polymer blending 2
1.3 Thermodynamics of polymer blends 3
1.4 Immiscible blends 5
1.5 Conclusions 9
1.6 Sources of further information and advice 10
1.7 References 11

2 **Characterization methods for high temperature polymer blends** 14
L. A. Utracki, formerly of the National Research Council, Canada

2.1 Introduction 14
2.2 High temperature polymer blends (HTPBs) 18
2.3 Methods of polymer characterization 23
2.4 Characterization of polymer blends 26
2.5 Characterization of HTPBs: chemical constitutions and molecular weights 29
2.6 Characterization of HTPBs: chemical-, thermal-, mechanical- and radiation-induced degradation 33
2.7 Stabilization of HTPBs 42
2.8 Challenges in blending polymers 45
2.9 Summary 49
2.10 Future trends 50
2.11 References 52
2.12 Appendix: Nomenclature 63

3 **Characterization of high temperature polymer blends for specific applications: fuel cells and aerospace applications** 70
L. A. Utracki, formerly of the National Research Council, Canada

3.1 Introduction 70
3.2 High temperature polymer blends (HTPBs) for membrane applications 73
3.3 Fuel cell (FC) membrane applications 74
3.4 Characterization of HTPBs for FC applications 79
3.5 Solar cell (SC) applications 86
3.6 Characterization of HTPBs for polymeric solar cells (PSCs) 93
3.7 Aerospace applications 99
3.8 Characterization of HTPBs for aerospace applications 104
3.9 Summary 109
3.10 References 110
3.11 Appendix: Nomenclature 123

4 **Thermodynamics of high temperature polymer blends** 130
M. T. DeMeuse, MTD Polymer Consulting, USA

4.1 Introduction 130
4.2 Blending miscible high temperature polymers 132
4.3 Poly (2,2′ (*m*-phenylene)-5-5′ bibenzimidazole) (PBI) blends 133
4.4 Polyimide blends 135
4.5 Liquid crystal polymer blends 135
4.6 Molecular composites 137
4.7 Conclusions 138
4.8 Sources of further information and advice 139
4.9 References 139

5 **Liquid crystal polymers (LCPs) as a reinforcement in high temperature polymer blends** 141
M. T. DeMeuse, MTD Polymer Consulting, USA and G. Kiss, Telecordia Technologies, USA

5.1 Introduction 141
5.2 Researching liquid crystal polymers (LCPs) 142
5.3 Liquid crystals 144
5.4 Polymer liquid crystals 145
5.5 Blends of isotropic and anisotropic thermotropic polymers 150
5.6 Processability of LCP/thermoplastic blends 152
5.7 Structure–property relationships of LCP blended materials 155

5.8 Commercial LCP blends 157
5.9 Conclusions and future trends 160
5.10 References 160

6 Polysulfones as a reinforcement in high temperature polymer blends 165
M. T. DeMeuse, MTD Polymer Consulting, USA

6.1 Introduction 165
6.2 Structure and properties of polysulfone 165
6.3 Issues in blending polysulfone with other high temperature polymers 166
6.4 Physical properties of polysulfone blends 168
6.5 Polysulfone/thermoset mixtures 169
6.6 Conclusions 171
6.7 Sources of further information and advice 172
6.8 References 172

7 Polybenzimidazole (PBI) high temperature polymers and blends 174
B. G. Dawkins, F. Qin, M. Gruender and G. S. Copeland, PBI Performance Products, Inc., USA

7.1 Introduction 174
7.2 Processing of polybenzimidazole (PBI) 177
7.3 PBI blends 179
7.4 PBI–polyetherketoneketone (PEKK) blends 183
7.5 PBI–polyetherimide (PEI) blends 188
7.6 PBI–polyaryletherketone (PAEK)–PEI blends 190
7.7 PBI–polyarylate (PA) blends 192
7.8 PBI–polysulfone (PS) blends 194
7.9 PBI–polyimide (PI) and PBI–polyamide-imide (PAI) blends 195
7.10 PBI–poly (bisphenol-A carbonate) (PC) and PBI–polybenzoxazole (PBO) blends 197
7.11 PBI–poly(4-vinyl pyridine) (PVPy) and other blends 198
7.12 PBI commercial products 199
7.13 PBI in high temperature applications 206
7.14 Future trends 208
7.15 Sources of further information and advice 209
7.16 References 209

Index *213*

This book is dedicated to the memory of the author of Chapters 2 and 3, Dr L. A. Utracki, who sadly died before the book was completed.

Contributor contact details

(* = main contact)

Editor and Chapters 1, 4 and 6

M. T. DeMeuse
MTD Polymer Consulting
10915 Arvind Oaks Ct.
Charlotte, NC 28277, USA

E-mail: kumquats88@aol.com

Chapter 5

M. T. DeMeuse*
MTD Polymer Consulting
10915 Arvind Oaks Ct.
Charlotte, NC 28277, USA

E-mail: kumquats88@aol.com

G. Kiss
Telecordia Technologies
1 Telecordia Dr.
Piscataway, NJ 08854, USA

E-mail: gkiss@telecordia.com

Chapter 7

B. G. Dawkins*, F. Qin, M. Gruender and G. S. Copeland
PBI Performance Products, Inc.
9800 D Southern Pine Blvd
Charlotte, NC 28273, USA

E-mail: bob.dawkins@pbiproducts.com

1
Introduction to high temperature polymer blends

M. T. DeMeuse, MTD Polymer Consulting, USA

DOI: 10.1533/9780857099013.1

Abstract: A review of definitions and the overall rationale for the production of high temperature polymer blends is provided. The discussion is divided essentially into two parts: miscible and immiscible blends. It is pointed out that one concern with miscible polymer pairs is that of processing in the miscible state. This phenomenon is dependent on the position of the phase separation temperature relative to the glass transition temperature of the polymer blend. In the case of immiscible blends, the issue of adhesion of the polymers is discussed. Finally, the need for better theoretical models for the prediction of miscibility in polymer blends is highlighted and discussed.

Key words: blends, high temperature, miscibility, immiscibility, glass transition, theoretical.

1.1 Introduction

Polymer blends are being used in an increasing number of industrial applications. Sectors ranging from the automotive to the aircraft industry have expressed an interest in developing these materials for specific applications. In fact, for some time, research in blends has been one of the biggest areas of polymer research, in both the industrial and academic world. Recently, requirements for materials in certain areas have become increasingly severe. Temperatures in excess of 200°C for times of hundreds of hours have become stated requirements for some materials. Particularly severe in this regard are requirements from the aircraft industry for engine components where high service temperatures for long periods of time are often normal.

The commercial activity in high temperature blends has been somewhat limited. As already alluded to, one of the major driving forces for higher temperature performance has involved military and aerospace applications. Other emerging application needs have developed, but at a slower rate. Continued needs for high temperature materials will be present in the aerospace as well as in the transportation area where under-the-hood components, friction and wear applications continue to require more demanding performance. Electric and electronic needs will also continue at a high rate of growth requiring higher temperature performance. Finally, high temperature polymer needs will emerge in advanced composites research, which often requires blend technology to optimize the performance of available materials. One approach to meeting these material requirements is to synthesize new polymers. Another method to tailor the properties of materials is through blending of two polymers. In this approach, the goal is to highlight the positive features of both materials while attempting to eliminate the negative features.

This book will address various aspects of high temperature polymers and highlight the advantages of producing such blends. Specifically, this first chapter will introduce the reader to general terminology and definitions and also provide background information on the various classes of high temperature polymer blends. For purposes of these discussions, high temperature polymers are generally defined as materials which have service temperatures of at least 175°C. This definition is arbitrary in nature and is governed by the fact that this is the temperature specified in many aerospace applications. Similarly, high temperature polymer blends are defined by the same use temperature. It should be noted that amorphous polymers that have a service temperature of 175°C have glass transition temperatures in the range of about 200°C, approximately 25°C higher than the actual service temperature. On the other hand, semicrystalline polymers have service temperatures which are largely defined by their melting temperatures. Most high temperature polymers are amorphous, but some are semicrystalline. In the case of these high temperature semicrystalline polymers, those with melting temperature less than 400°C are desirable. Above that temperature, degradation starts to compete with melting.

1.2 General principles of polymer blending

The physical state of mixing which is present in polymer blends can be categorized in three ways: miscible, partially miscible or immiscible. A miscible polymer blend is a homogeneous, single-phase material. In many respects, a miscible blend behaves as if it is a single polymer. If the two components in the blend are miscible only within a certain composition range, the polymers are deemed partially miscible. An immiscible blend contains two distinct phases and is heterogeneous in nature.

The philosophy of blending two polymers is to produce a material which has properties that are tailored to a certain performance level. Since the morphology and resultant physical properties are controlled to a large extent by the miscibility and phase behavior, it is quite easy to understand why there is so much interest in controlling the phase structure in blends.

In fact, this topic has evolved into a central area of polymer research during the last 40 years. One of the first ideas, that as a rule polymer blends are immiscible, needs to be reevaluated due to the increasing number of miscible or partially miscible polymer pairs reported in the literature (see, for example, Paul and Newman, 1979).[1] Despite this high level of activity, much of the work remains based on art and intuition rather than on science. Most of the work performed on high temperature polymer blends has involved the definition of miscible polymer pairs and their phase separation characteristics.[2–6] Not much work has been done to predict miscible pairs. In fact, this is an open area of research in the entire area of polymer blends.

Present theoretical models of polymer blends generally focus on the extreme cases of either random coil mixing or rigid rod mixing. For treating random coil mixtures, the Flory–Huggins formalism[7] is usually used to describe the free energy of mixing. In this formalism, miscibility is largely dominated by the enthalpy part of the free energy. On the other hand, Flory[8,9] has treated the mixing of rigid rods with random coils, the concept upon which molecular composites are based. In these systems, the two polymers being mixed have vastly different molecular conformations. Phase separation is easily introduced in such systems. It should be recognized that this separation is based solely on entropic effects.

Two extreme cases have, then, been established in the literature. Random coil/random coil mixing is controlled by enthalpic effects, while the mixing of rigid rods and random coils is dominated by entropic effects. However, since most high temperature polymers contain heterocyclic units, they possess restricted rotation in their backbone. In fact, it is precisely this restricted rotation which leads to the high glass transition temperature of these polymers. Thus, these high temperature polymers can be regarded as rigid, or at least semi-rigid, in overall conformation. An adequate theoretical model is still to be developed for the mixing of such polymers.

1.3 Thermodynamics of polymer blends

A miscible polymer blend is a homogeneous, single-phase material. Various experimental methods, many of which will be discussed in Chapter 2, have been developed for defining such a structure in a mixture of polymers. Among these, a single glass transition temperature (T_g) is the most widely accepted criteria for miscibility. It is generally assumed that polymer blends which display a single T_g are mixed at the molecular level. Based on these criteria, various models[10,11] have been reported in the literature for the description of the relationship between T_g and the blend composition.

A necessary condition for two polymers to form a miscible blend is that the free energy of mixing, ΔG_{mix}, should be less than zero, where ΔG_{mix} is given by the following equation:

$$\Delta G_{mix} = \Delta H_{mix} - T\,\Delta S_{mix} \tag{1.1}$$

where ΔH_{mix} is the enthalpy of mixing, T is the temperature and ΔS_{mix} is the entropy of mixing. Obviously, one way for ΔG_{mix} to be negative in Equation [1.1] is for ΔH_{mix} to be negative. This is generally accomplished through specific interactions, such as dipole-dipole or hydrogen bonding.

Simply having a negative free energy of mixing does not guarantee phase stability. The necessary condition for phase stability in a blend composition x at a fixed temperature and pressure is that the second derivative of the free energy with respect to composition is greater than zero, i.e.,

$$\frac{\partial^2 \Delta G_{mix}}{\partial x^2} > 0. \tag{1.2}$$

Without satisfaction of this condition, the miscible system is thermodynamically unstable and phase separation will occur.

One of the most common models for treating the free energy of mixing in polymer blends is the Flory–Huggins treatment.[7] That model assumes that the free energy of mixing, ΔG_{mix}, is given by:

$$\Delta G_{mix} = \frac{\Theta_1}{N_1}\ln\Theta_1 + \frac{\Theta_2}{N_2}\ln\Theta_2 + \Theta_1\Theta_2\chi_{12} \tag{1.3}$$

where Θ_1 and Θ_2 are the volume fractions of polymers 1 and 2, respectively, N_1 and N_2 are the degrees of polymerization of polymers 1 and 2, and χ_{12} is called the 'chi' parameter and is used to describe the enthalpic interaction between polymers 1 and 2. This, as originally developed, is strictly applicable only to polymers which have random coil conformations. It is also assumed that the entropic and enthalpy contributions to the free energy of mixing are completely independent of each other.

Based on the rigid conformation of many high temperature polymers, it is expected that the criteria for miscibility of such polymers will be even stricter than for random coil polymers. However, several high temperature miscible polymer blends have been discovered and reported in the literature[2–6,12–16] which are based on poly(2,2´ (m-phenylene)-5-5´ bibenzimidazole) (PBI), the chemical structure of which is shown in Fig. 1.1. Chapter 7

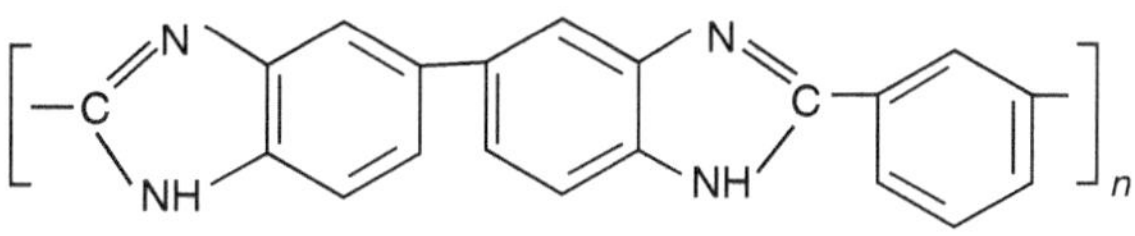

1.1 Chemical structure of poly(2,2′ (m-phenylene)-5-5′ bibenzimidazole) (PBI).

discusses these blends in detail but it is to be noted that the fundamental reason for the observation of miscibility in these systems is the occurrence of functional groups in the polymer backbone which can interact with the functional groups present on the backbone in the other polymer in the mixture. Thus, miscibility in these systems is an example of a specific interaction which leads to a negative enthalpy of mixing.

One of the issues with miscible polymer blends in general is the temperature range of their miscibility. If such materials are to be processed in the melt state as miscible blends, there must be a temperature window between the glass transition temperature and the phase separation temperature which is large enough for melt processing. If this processing window is too small, the blends will either phase separate or have too high a viscosity for melt processability.

In the case of the PBI blends discussed above, phase separation takes place a few degrees above the glass transition temperature for most blend compositions.[17] This can be seen from the phase diagram for PBI/Ultem poly (ether imide) blends, as shown by Karasz and MacKnight.[17] This suggests that the miscibility which is observed is a metastable phenomenon which is controlled by kinetic factors. The observed phase separation also indicates that it will be difficult to melt process the blends in the miscible state, due to the reasons already discussed.

There have been several miscible high temperature polymer pairs defined in the literature. Several of these pairs are miscible from solution but are immiscible when processing is attempted in the melt state. These results indicate that the blends phase separate when heated above their glass transition temperature. This further shows that kinetic factors as well as thermodynamic factors are important in the observed miscibility. Also, the role of the solvent in the observed miscibility needs to be better understood. One of the current technical challenges is to widen the temperature range between the glass transition temperature of the blend and its phase separation temperature, to allow miscible blends to be processed in the melt state.

1.4 Immiscible blends

In this section, three types of immiscible blend will be briefly introduced. The first will be molecular composites, which are generally accepted as partially miscible blends but are not truly miscible at the molecular level. The second type of immiscible blend involves mixtures of two thermoplastic polymers, and a discussion is provided of the features that govern the observed structure. The final type will be mixtures of thermosets with high temperature thermoplastic polymers.

1.4.1 Molecular composites

The concept of molecular composites was first proposed by Flory in the late 1970s. These systems are based on the mixing of a rigid rod polymer and a random coil polymer, vastly different molecular conformations. Flory[8,9] predicted that phase separation is easily induced in these systems and is based solely on entropic effects.

Experimental work on molecular composites has focused on attempting to kinetically delay the phase separation with a desirable morphology before the thermodynamics leads to complete immiscibility. Most of this work was initially performed as part of a program sponsored by the United States Air Force. For example, poly (p-phenylene terephthalamide) (PPTA) and poly (p-phenylene benozbisthiazole) (PBT) have been successfully dispersed in a Nylon 66 matrix.[18–22] Kyu *et al.*[23] have reported that there are interactions present in a PPTA and Nylon 6 system and that phase separation can be thermally induced in molecular composites based on these two polymers. Thermally induced phase separation has also been observed in the PBT/Nylon 6 system, when the melting temperature of the Nylon component is reached. Finally, Moore and Mathias[24] reported a unique method for the preparation of molecular composites using an *in situ* polymerization process in which the anion of the PPTA was used as the initiator for the anionic polymerization of acrylamide in the formation of a Nylon 3 matrix.

The molecular composites discussed thus far involved processing of the two polymers from a common solvent. One of the major challenges of this approach is the identification of a proper solvent that can be used. In addition, most of the suitable solvents are quite corrosive in nature. That fact, as well as the control of phase separation and, hence, the final properties have largely hindered further widespread industrial development of these materials.

Initial applications of molecular composites are expected to be in the aerospace area. Matrix materials based on molecular composites which are reinforced with carbon fiber are expected to be a key area of interest. The use of a molecular composite as a matrix for advanced composites should provide a significant increase in the modulus and strength over that obtainable with conventional thermoplastics and thermosets. Electronic applications are another area where molecular composite commercialization should continue to emerge.

1.4.2 Immiscible thermoplastic blends based on liquid crystal polymers

It has been established in the literature[25] that one of the key parameters which established the morphology and, hence, the physical properties of blends which are based on two immiscible polymers is the viscosity ratio for the two component polymers. The ratio can usually be controlled through parameters such as the molecular weight of the polymers and the melt processing temperature. Through that approach, it is possible to observe a variety of different morphologies with the same two immiscible polymers.[26]

Nowhere is that fact more apparent than in blends of liquid crystal polymers (LCPs) with other thermoplastic polymers. It is not the intent of the present chapter to completely review all the work done in the area of blends which contain LCPs. Instead, that will be the focus of Chapter 5. The objective of the present discussion is to make general observations and show how those general phenomena may be applicable to other high temperature polymer blends as well.

In these types of blends, the size, shape and distribution of the LCP phase depends on many factors, such as the blend composition, the processing conditions, the viscosity ratio of the component polymers at the shear rate which is being used in the processing, and the rheological characteristics of the thermoplastic matrix polymer. This observation is present in the work of Acierno *et al.*[27] who demonstrated that different morphologies could be observed in the same LCP/polycarbonate blend simply by varying the processing temperature. The observed morphological differences were attributed solely to different viscosity ratios at the different processing temperatures.

Several reports have appeared in the literature which discuss the lowering of the viscosity of a thermoplastic polymer with the addition of a small amount of the LCP component. These studies have been performed on a variety of polymers including polyamide,[28] poly (ether imide),[29] poly (ether sulfone)[30,31] and polycarbonate.[32–34] All of these studies reported a lowering of the viscosity of the traditional thermoplastic polymer with the addition of various small amounts of the LCP component.

This result is interpreted as being due to a lubricating effect of the LCP on the melt. This is because the groups of LCP molecules, called domains, slide past each other resulting in a lubrication of the polymer melt.

This lowering of the melt viscosity allows the LCP to act as a processing aid for the conventional thermoplastic. Cogswell *et al.*[35–37] concluded that, in order for this effect to be realized, the temperature range at which the conventional polymers are melt processed should overlap the temperature range in which the LCP forms an anisotropic melt. In that work, as in the other previously reported studies, the melts had lower viscosities than the pure thermoplastic polymers and, therefore, the processing temperature could be lower. The advantages of the subsequent reduction in the processing temperature include reduced energy consumption and less degradation of polymers that are sensitive to high temperatures. Also, a lowering of the viscosity allows for easier filling of large and complex molds in injection-molding applications.

The other principal effect observed in blends of LCPs with thermoplastic polymers is the utilization of the LCP as a reinforcement for the more flexible polymer. In numerous studies reported in the literature[38–53] improved mechanical properties in the blends have been observed. Most of these studies have attempted to explain the changes in the mechanical properties in terms of the morphology of the LCP domains in the blends. The most widely used experimental technique to study the morphology has been the scanning electron microscope (SEM).

The fact that LCPs can act as reinforcing agents in a blend has led some workers to model the mechanical behavior of the blends using theories of composites. Thus, Kohli *et al.*[47] showed that the modulus of highly drawn melts can be treated effectively by a simple rule of mixtures. That is, the modulus of a blend is given by

$$M_{\text{blend}} = \upsilon_1 M_1 + \upsilon_2 M_2 \qquad [1.4]$$

where υ_1 and υ_2 are the volume fractions of the two components in the blend and M_1 and M_2 are the corresponding modulus values of the two polymers. These same workers showed that the moduli values of samples in which the LCP forms spherical particles can be treated using an inverse rule of mixtures.

One of the main drawbacks of using LCPs as a reinforcement is the poor adhesion to the matrix polymer. This lack of adhesion between two immiscible polymers is a general phenomenon and appears to be common to all blends which contain LCPs. An approach to address this problem was proposed by Akkapeddi *et al.*[54] They proposed blending thermotropic oligomers and isotropic polymers in the presence of a particulate material such as talc or silica. The purpose of the particulate is twofold. First, it appears to reduce the phase separation between the thermotropic oligomer and the matrix polymer. Second, its presence helps improve the dispersion of the oligomer in the polymer. The final blend has an increased tensile modulus, tensile strength and abrasion resistance compared to blends in which the particulate was not used.

Summarizing the work on immiscible blends which contain LCPs, two interesting phenomena have been demonstrated in the literature. First, the addition of an LCP to a thermoplastic polymer can be used to lower the melt viscosity, allowing for lower processing temperatures and, perhaps, easier mold filling. The second phenomenon which has been described is the use of LCPs as reinforcing agents. This observation has been made with several thermoplastic polymers, and the processing conditions for obtaining this morphology have been defined.

The biggest technical issue which has prevented more widespread use of these blends is the lack of adhesion between the LCP and the matrix polymer. Additional work needs to be focused on solving this problem before the full potential of these blends can be realized. Effective ways to compatibilize the two polymers need to be further developed. This is a general effect of blends of two immiscible high temperature polymers, and additional work is needed in this area.

1.4.3 Semi-interpenetrating networks

The idea behind semi-interpenetrating networks (semi-IPNs) is to combine the processability of thermoplastic polymers with the high temperature performance of cross-linked thermosetting materials. Such mixtures should possess the desirable features of both types of materials. Most of the semi-IPNs contain continuous phases, and the components are immiscible at the molecular level.

St. Clair *et al.*[55,56] used the concept of sequential semi-IPNs to produce two unique IPNs which are based on acetylene-terminated imide oligomers and thermoplastic polyimides, particularly a material designated LARC-TPI. Pater and others[57–59] used the concept of simultaneous semi-IPNs to make semi-IPNs which are based on PMR-15 mixed with such materials as LARC-TPI, NR150B2 and Thermid 600 polyimides. Other examples of semi-IPNs which fit into the category of high temperature materials include thermoplastic modified bismaleimides (BMI). As one example, BMI has been mixed with condensation polyimides such as PI20280[60] or with terminated poly (arylene ether ketone) oligomers.[58] The observed miscibility in semi-IPNs which are based on thermoplastic BMIs was found to be further improved when a BMI and thermoplastic both prepared from aromatic diamines were blended.[61,62]

1.5 Conclusions

Several common themes have emerged in all of the systems that have been discussed. The first is related to the high temperature involved in processing high temperature blends. This high temperature means that possible degradation of the polymers is always an issue. Such degradation may often compete with this processing of the polymer mixture. A second issue is the control of the phase separation in the blends. Sometimes, neither complete miscibility nor immiscibility is desired in blend systems, but instead some intermediate situation. Thus, complete miscibility may not be the optimum situation, but enough adhesion between the two phases must be present to obtain a desired physical property profile. The mechanism of the transformation from a miscible to immiscible system and vice versa is presently not well understood. The thermodynamics of phase separation may be understood, but the kinetics is still to be developed.

The morphology which results from the phase separation of a binary polymer mixture during a transition across the phase boundary is an area of considerable interest, but is one hardly addressed in the case of high temperature polymer mixtures. The resultant rheology of phase separated polymers will be a function of the morphology. Prediction of the rheological changes which accompany the morphological changes is an area where further research is needed.

In addition, in blends which contain two-phase separated polymers, adhesion of the two phases is of primary importance for utilizing the blend. Additional research needs to be done to understand the features that provide compatibilization in high temperature polymer blends and the optimum structure of compatibilizing agents for such blends. An example of this has been presented in the case of blends which contain liquid crystal polymers, but the same point is true of all blends based on immiscible high temperature polymers.

Better theoretical models and predictive tools need to be developed which can be used to understand the structural features in polymers which lead to miscibility. Presently, most of the information obtained in this regard has been done through trial and error. Then, when a miscible system is found, efforts are expended to rationalize the observed miscibility. A different approach would be to develop theoretical schemes which can predict miscibility beforehand and, then, experimentally produce such systems. Unfortunately, the present computer technology does not yet adequately allow for the incorporation of explicit molecular details in such models. When such developments become available, a major step in the development of high temperature polymer blends will be achieved.

1.6 Sources of further information and advice

There are several excellent books available which discuss polymer blends in general, including the thermodynamics of phase separation.[1,62,63] The interested reader is referred to any of those texts to gain additional insights into polymer blends. A good reference to learn more about high temperature polymers is provided by Hodd,[64] whose article provides a background on many high temperature polymers as well as trends in the applications of those materials.

1.7 References

1. D.R. Paul and S. Newman in *Polymer Blends*, Academic Press, New York, New York (1979).
2. L. Leung, D.J. Williams, F.E. Karasz and W.J. MacKnight, *Polym. Bull.*, **16**, 1457 (1986).
3. G. Guerra, S. Choe, D.J. Williams, F.E. Karasz and W.J. MacKnight, *Macromolecules*, **21**, 231 (1988).
4. P. Musto, F.E. Karasz and W.J. MacKnight, *Polymer*, **30**, 1012 (1989).
5. C. Choe, W.J. MacKnight and F.E. Karasz, *Polym. Mater. Sci. Eng.*, **59**, 702 (1988).
6. S. Stankovic, G. Guerra, D.J. Williams, F.E. Karasz and W.J. MacKnight, *Polym. Commun.*, **29**, 14 (1988).
7. P.J. Flory, *Principles of Polymer Chemistry*, Cornell University Press, Ithaca, New York (1953).
8. P.J. Flory, *Macromolecules*, **11**, 1138 (1978).
9. P.J. Flory, *Macromolecules*, **11**, 1141 (1978).
10. L.A. Wood, *J. Polym. Sci.*, **28**, 319 (1958).
11. P.R. Couchman and F.E. Karasz, *Macromolecules*, **11**, 117 (1978).
12. S. Makhija, E.M. Pearce and T. Kwei, *Polym. Eng. Sci.*, **30**(13), 798 (1990).
13. M. Jaffe, paper presented at 'Recent Advances in Polyimides and Other High Temperature Polymers', ACS, San Diego, January (1990).
14. D.S. Lee and G. Quinn, *Polymer*, **21**(9), 751 (1989).
15. C.A. Arnold, D. Chen, Y.P. Chen, J.D. Graybeal, R.H. Bott, T. Yoon, B.E. McGrath and J.E. McGrath, *Polym. Mater. Sci. Eng.*, **59**, 934 (1988).
16. T.S. Chung and P.N. Chen Sr., *J. Appl. Polym. Sci.*, **40**, 1209 (1990).
17. F.E. Karasz and W.J. MacKnight, paper presented at 2nd Topical Conference on Emerging Technologies in Materials, San Francisco, CA, November (1989).
18. M. Takayanagi, T. Ogata, M. Morikawa and T. Kai, *J. Macromol. Sci. Phys.*, **B17**(4), 176 (1978).
19. M. Takayanagi, T. Ogata, M. Morikawa and T. Kai, *J. Macromol. Sci. Phys.*, **B22**, 231 (1983).
20. M. Takayanagi, *Pure Appl. Chem.*, **55**, 819 (1983).
21. M. Takayanagi and K. Gotto, *Polym. Bull.*, **13**, 35 (1985).
22. W.F. Hwang and D.R. Wiff, US Patent 4,631,318 (1986).
23. T. Kyu, T.I. Chen. H.S. Park and J. White, *J. Appl. Polym. Sci.*, **37**, 201 (1989).
24. D.R. Moore and L.J. Mathias, *J. Appl. Polym. Sci.*, **32**, 6299 (1986).
25. E.W. Okoroafor, J.P. Villemaire and J.F. Agassant, *Polymer*, **33**(24), 5264 (1992).
26. V. Everaert, L. Aerts and G. Groenickx, *Polymer*, **40**(24), 6627 (1999).
27. D. Acierno, E. Amendolo, C. Cartagna, L. Nicolais and R. Nobile, *Mol. Cryst. Liq. Cryst.*, **153**, 533 (1987).
28. A. Siegmann, A. Dagan and S. Kenig, *Polymer*, **26**, 1325 (1985).

29. S. Swaminathan and A.I. Isayev, *Proc. ACS Div. Polym. Mat.*, **57**, 330 (1987).
30. S.G. James, A.M. Donald and W.A. MacDonald, *Mol. Cryst. Liq. Cryst.*, **153**, 491 (1987).
31. M.F. Froix, M. Park and N. Trouw, US Patent 4,460,736 (1984).
32. A.I. Isayev and M.J. Modic, *Polym. Composites*, **8**, 158 (1987).
33. T.M. Malik, P.J. Carreau and N. Chapleau, *Polym. Eng. Sci.*, **29**, 600 (1989).
34. A. Kohli, N. Chung and R.A. Weiss, *Polym. Eng. Sci.*, **29**, 573 (1989).
35. F.N. Cogswell, B.P. Griffin and J.B. Rose, US Patent 4,386,174 (1983).
36. F.N. Cogswell, B.P. Griffith and J.B. Rose, US Patent 4,433,083 (1984).
37. F.N. Cogswell, B.P. Griffin and J.B. Rose, US Patent 4,438,236 (1984).
38. T. Chung, *Plast. Eng.*, **43**, 39 (1987).
39. T. Chung, *Proc. Ann. Tech. Conf. Soc. Plast. Eng.*, **33**, 1404 (1987).
40. R.A. Weiss, W. Huh and L. Nicolais, *Polym. Eng. Sci.*, **27**, 684 (1987).
41. G. Kiss, *Polym. Eng. Sci.*, **27**, 410 (1987).
42. K.G. Blizzard and D.G. Baird, *Polym. Eng. Sci.*, **27**, 653 (1987).
43. R.A. Weis, W. Huh and L. Nicolais, International Conference of Liquid Crystal Polymers, Bordeaux, France (1987).
44. R.A. Weiss, W. Huh, L. Nicolais and A. Kohli, Conference on Emerging Technologies in Materials, AIChE (August 1987).
45. R.A. Weiss, W. Huh, L. Nicolais and P. Yanisko, *Proc. Reg. Tech. Soc. Plast. Eng.*, p. 267 (1987).
46. R.A. Weiss, W. Huh and L. Nicolais, in *High Modulus Polymers*, A.E. Zachariades and R.S. Porter, eds., Marcel Dekker, New York, p. 145 (1988).
47. A. Kohli, N. Chung and R.A. Weiss, *Polym. Eng. Sci.*, **29**, 573 (1989).
48. N. Wellman, G. Hussman, A.K. Kulshrestha, T. Helminiak, D. Wiff, C. Benner and W.F. Hwang, *ACS Organic Coat, Plast. Chem.*, **43**, 783 (1980).
49. O.A. Nehme, C.A. Gabriel, R.J. Farris, E.L. Thomas and M.F. Malone, *J. Appl. Polym. Sci.*, **35**, 1955 (1988).
50. E.G. Joseph, G.L. Wilkes and D.G. Baird, *ACS Polym. Preprints*, **25**(2), 94 (1984).
51. R.A. Ramanathan, K.G. Blizzard and D.G. Baird, *Proc. Ann. Tech. Conf. Soc. Plast. Eng.*, **33**, 1399 (1987).
52. M. Amano and K. Nakagawa, *Polymer*, **28**, 263 (1987).
53. R.A. Weiss, A. Kohli, N.S. Chung and D. Dutta, unpublished results.
54. M.K. Akkapeddi, T. DeBona, H.L. Li and D.C. Prevorsek, US Patent 4,611,025 (1986).
55. A.O. Hanky and T. St. Clair, *SAMPE J.*, July/Aug., 40 (1985).
56. A.H. Egli, L.L. King and T. St. Clair, 18th International SAMPE Technical Conference, (October 1986).
57. R. Pater and C. Morgan, *SPE ANTEC Proceedings*, **1639** (1988).
58. R. Pater, paper presented at the RETEC Technical Conference, Los Angeles, CA, November (1988).
59. R. Pater and C.D. Morgan, *SAMPE J.*, **24**(5), Sept./Oct. (1988).
60. P.A. Steiner, J.M. Browne and M.T. Blainand McKillen, *SAMPE J.*, March/April (1987).
61. C.A. Arnold, D. Chen, Y.P. Chen, J.D. Graybeal, R.H. Bott, T. Yoon, B.E. McGrath and J.E. McGrath, *Polym. Mater. Sci. Eng.*, **59**, 934 (1988).

62. O. Olabisi, L. Robeson and M.T. Shaw, *Polymer-Polymer Miscibility*, Academic Press, New York (1979).
63. L.A. Utracki, *Polymer Alloys and Blends: Thermodynamics and Rheology*, Hanser Publishers, New York (1990).
64. K. Hodd, *Trends. Polym. Sci.*, **1**(5), 129 (1993).

2
Characterization methods for high temperature polymer blends

L. A. UTRACKI, formerly of the
National Research Council, Canada

DOI: 10.1533/9780857099013.14

Abstract: This chapter provides information about characterization methods for examining the structure and properties of blends that contain at least one high temperature polymer. Methods are described that probe both the phase structure of such blends as well as techniques that define and measure the properties of blends. Relations between the determined phase structure and resultant properties are also discussed.

Key words: high temperature polymer blends, polymer blend degradation, polymer characterization, blending polymers.

2.1 Introduction

The first patent relating to polymer alloys and blends (PAB) was granted to Parkes in 1846.[1] It described blending two types of polyisoprene, amorphous natural rubber (NR) and semi-crystalline gutta-percha (GP). Neither the amorphous, sticky, highly deformable natural rubber nor its isomer, crystalline gutta-percha, was useful, but blending them created a rubber with superior processability and properties. Since then the blending of polymers has become widespread. Blending benefits the manufacturer as well as the consumer. The former enjoys improved production flexibility, better processability, product uniformity, and scrap reduction, while the latter benefits by having tailor-made materials with a full set of desired properties. There are a number of books which review the fundamental aspects of PABs,[2,3] the evolution of commercial PABs[4,5] and the commercial blends available in the market.[6] There are also major reference works on PAB, such as those by Utracki and Isayev.[7,8]

While the early plastics were bio-based (e.g., rubbers, cellulose, starch, urea-formaldehyde) at present 95 wt% of plastics are synthetic, consuming 4–8% of the available petroleum feedstock. Packaging uses the most plastics (32%), followed by building construction (20%), consumer goods (18%), electrical and automotive engineering (each 9%), agriculture (6%) and other sectors (the remainder). World plastics production to meet this need has followed an exponential curve clearly influenced by political events, for example, slow-down during periods of depression (e.g., in the 1930s) and acceleration during wars (e.g., during WWII).[5] World plastics production in 1900 was about 30 thousand tons (kt), increasing to 300 million tons (mt) by the year 2010. Figures 2.1 and 2.2 show the growth after 1960, extrapolated to 2020.[9]

Dozens of blends are in the market. Since synergism is rare, new blends are designed by adding components to achieve the desired performance characteristics. Historically blend complexity has tended to increase; for example, while initially two-component blends were sufficient (e.g., PPE + PS or HIPS) nowadays blends with seven polymers + compatibilizers are available (e.g., PPE + PS or HIPS, PC, ASA, SAN, SEBS and PTFE[10]) or alternatively PPE with either PS, SEBS or HIPS, PC, PEST, ABS, PMMA and copolymer of alkyl and glycidyl methacrylate.[11]

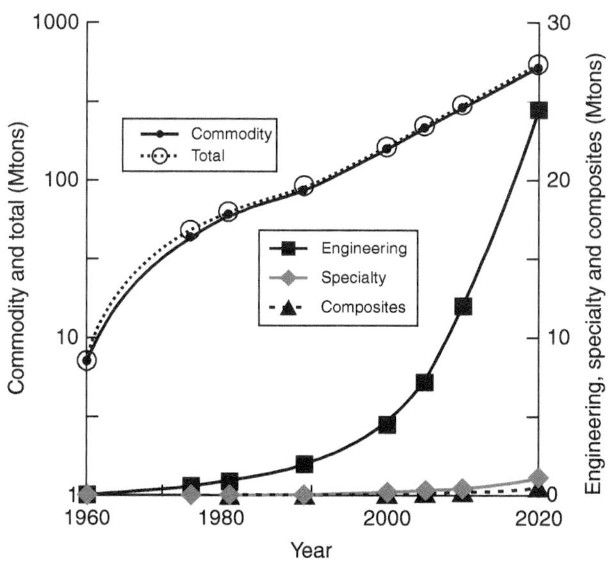

2.1 World plastics production in million tons, extrapolated to 2020. Data from reference 9.

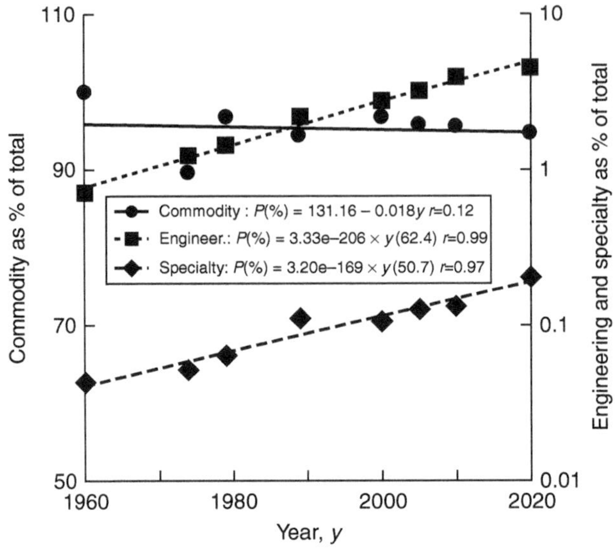

2.2 The plastics production data from Fig. 2.1 converted to percent of total production. *P*, world plastics production.

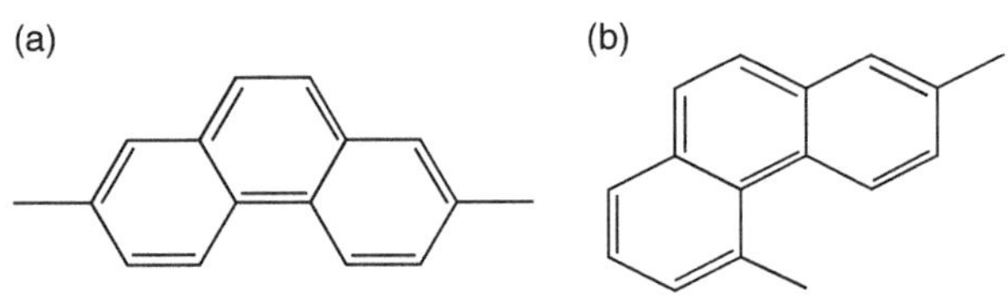

2.3 Example of rigid-rod (a) and kink-forming (b) monomeric structures.

Physical modification usually leads to molecular orientation of high molecular weight semicrystalline polymers, such as PE, PP, or PARA produced in the form of fibers or tapes. Examples include Dyneema®, Innegra™ and Kevlar.[12–14] Since high performance is often associated with a rigid molecular structure, these materials are excellent for applications such as reinforcing composites.[15] Chemical modification also opens the way to new, better performing blends. For example, Solvay Advanced Polymers deposited a patent application for a new PPE obtained by condensation *para*-substituted polyphenylene and kink-forming arylene units (see Fig. 2.3). By combining two types of polyarylene macromolecules the inventors obtained materials with an exceptional set of mechanical properties and flame resistance.[16] In essence, the invention describes blends of two high performance copolymers containing substituted or not substituted *p*-phenylene with unsubstituted *m*-phenylene, i.e., two types of rigid rod and flexible chain statistical segments that are miscible in the amorphous phase and self-reinforced by forming high aspect ratio rigid-rod crystals.

Copolymers with different concentrations of linear and kinked monomers can be melt-blended with each other or with other high temperature polymers (HTPs) into a variety of grades optimized for aerospace applications, surgical and medical devices, silicon wafer processing in electronics, military and defense applications, membranes for separation of gases, liquids and solids, etc. These self-reinforcing polyphenylenes (SRPs) show high strength and stiffness, outstanding mechanical performance, flame resistance, chemical resistance, surface finish, low thermal expansion and H_2O absorption. As an example, Table 2.1 lists a variety of PA blends. Similar tables can be constructed for any polymer in Table 2.2.[2–8]

Table 2.1 PA blends with specialty polymers[17]

Composition	Comment
1. PA/PSF blends	
PSF with PA-11	For improved impact properties
PA/PSF with poly(sulfone-g-lactam)	Processability and mechanical properties
PA-6/PSF	Processability, thermal and mechanical properties
PARA/PES	HDT = 172°C and mechanical properties
PA-46/PES	Mechanical and thermal properties
PARA/PES	Mechanical and thermal performance
PAES or PES blended with PA-6T6	High heat resistance and stiffness
2. PA/PPS blends	
PARA with a small amount of PPS	Moldability, HDT, and impact strength
PA-46 with PPS	Impact and mechanical properties, HDT, heat and chemical resistance
PPS with polyphthalamide (PPA)	Heat and chemical resistance, HDT, mechanical properties
PPS with PPA and GF	Mechanical properties
PPS/PA with EPDM and MA	Compatibilized-*cum*-impact modified
PPSS with either PA or PEST	Mechanical properties, HDT, dimensional and chemical stability
PPS, either PA-66 or PA-MXD6 and Mg(OH)2	Tensile strength, arc tracking, and heat resistance
3. PA/*phenoxy* blends	
PA-6 with *phenoxy*	Excellent ESCR
PA-66 with *phenoxy* and SEBS	Tensile and impact strength
PA with PEST, *phenoxy* and MBS	Tensile and impact properties
4. PA/PEA blends	
PA-12 with PEA	Tough, flexible, heat and hot oil resistant
PEBA/PA and (ABS, MBS, NBR, SBR or EPDM)	Impact strength at low-T, *Rilsan*™

(*Continued*)

Table 2.1 Continued

Composition	Comment
5. PA/PAI blends	
PAI with PA-66 or PARA and inorganic filler	Moldings with high mechanical performance
PI, PA, PAI or PI + PA and aromatic PI	Miscible blends for selective permeation
6. PA/PEI blends	
PEI with PA or PEST	Processability
PEI with PA-6 and PEI-b-PA	Impact strength
PEI with PA-6 or PA-66	Compatibilized by nonyl-phenolic
PEI with PA-12	Reduced shrinkage and water absorption
Copolyesteretherimide, PEEI *Lomod*™, blended with either PA or PARA	Thermal aging behavior

2.2 High temperature polymer blends (HTPBs)

As shown in Table 2.3, plastics can be grouped into four categories according to continuous use temperature (CUT):

- Commodity (CUT ≤ 75 °C): a category which includes high and low density polyethylenes (HDPE and LDPE), polypropylenes (PP), styrenics (PS), acrylics (PMMA), and polyvinylchloride (PVC). For recycling these resins are identified with a numbered triangle, namely 2 = HDPE, 3 = PVC, 4 = LDPE, 5 = PP, 6 = PS. The category 1 denotes engineering (PET) and all others are marked as 7.
- Engineering (CUT ≤ 140 °C): a category comprising polyamides (PA), thermoplastic polyesters (PEST including PET, PBT), polycarbonate of bisphenol-A (PC), polyoxymethylene (POM or acetal), and poly(2,6-dimethyl-1,4-phenylene oxide), better known as polyphenylene ethers (PPE)
- Specialty (CUT 140 to 240 °C): including, for example, poly-aromatic amides (PARA), polyphthalamide (PPA), polysulfones (PSF, PSU), polyphenylene sulfone (PPSU), polyethersulfone (PES), polysulfides (PPS), liquid crystal polymers (LCP), etc. These polymers show high mechanical performance.
- An 'ultra' class of high performing polymers (CUT > 240 °C): including polybenzimidazole (PBI), polyetheretherketone (PEEK), polyimide, polyamide-imide, polyether-imide (PI, PAI, PEI), high temperature sulfone (HTS), self-reinforced polyphenylene ether (SRP), etc.

Table 2.2 High temperature polymer blends (HTPBs) arranged in order of increasing 'continuous use temperature' (CUT)

Polymer	Examples	Temperature (°C)			Blends with	Comment
		CUT	T_m	T_g		
Polyoxymethylene	POM	85–115	160–175	–80	POM, PEST, PPE, PC	Delrin®
Polyphenyleneether	PPE (PPE/PS)	104	257	205–215	PEST, PC, PA, PARA, PCT, POM	With PPE or styrenics
Polyesters	PET, (PBT, PC)	110, (118)	257, (219)	70, (31)	PBT, PEN, PAr, POM, PCT, PEA, PPE/PS, PC	Valox®
Polyamides	PA-6, (PA-66)	110, 0(99)	220 – 225, (255 – 264)	52, (65)	PPE, PA, PARA, POM, LCP, PET, PBT, PAr, PEB, PC	Nylon, Zytel® (aliphatic)
Polycarbonate	PC	116		147–149	PET, PBT, PCT, PCTG, PAr, PPE/PS, POM	Bis-phenol-A polycarbonate, PC
Polyphthalamide	PPA	120–185		310	PA, PPS, PP, PARA, PC, LCP	Amodel®
Polyarylates	PAr	130		188	PEST, PARA, LCP, PA, PBI, PC, PEI, PEK, PEST, Phenoxy, PMB, PPE, PPS	With heteroatom in the main chain
Polyarylene sulfide	PPS	170	280– 288	194	PPE, PC, PTFE, PSF, PAr, PARA, PI, PAI, PEK, LCP, PAES, PPSS, PDMS, FEP, PCTFE, PTFE, PVDF	Diverse aromatic; Rayton®
Aromatic (amorphous) polyamides, PARA*	PA-6I, (PA-6I6T),	[180]	–, (310), [310]	120, (90), [127]	POM, PEST, PA, PAI, PPS, PES, PPhA, PAr, PBI	Excluding LCP-type

(*Continued*)

Table 2.2 Continued

Polymer	Examples	Temperature (°C)			Blends with	Comment
		CUT	T_m	T_g		
Polyphenylene-terephthalamide[†]	PPTA, PMTA	180	310	90	PBI	Flame resistant coatings
Polysulfone, PSU, polyphenylsulfone,	PPSU (PSF), [PES], PSO	190, (140 – 150), [204]		190, [230]	PC, PPE, PA, PEST, PI, PAI, PEI, PEK, PEEK, PTFE, PEA	Mindel®, Eviva® Udel®, Radel®,
Liquid crystal polymers	LCP	220 – 271	300 – 350	ca. 120	PA, PPTA, PARA, PAEK, PPE, PI, PAI, PEI, PPS, PEST, PVDC, PVDF, PVF, PSF, LCP, fluorinated polymers	Xydar®, diverse others
Polyaryletherketones, PAEK	PEEK	240 –310	334	143	PES, PPS, PI, PAI, PEI, PA, LCP, PSF, fluorinated polymers	Victrex®
Polyimides	PAI (PEI)	250 – 275, (165 – 170)		279, (220)	PA, PARA, PEST, PC, PPE, PPS, PSF, PEEK, LCP (PC, PA, PSF, PEC, PAr, PEST, LCP, PPS, PEEI)	Ultem®
Thermoplastic polyimide	TPI	230 [465]	– [388]	≤ 311 [250]	PEI, LCP, PTFE, PAI, PPE, PARA, PPS, PSF, PEEK, PI	Extem®, Aurum®
Polybenzimidazole[‡]	PBI	435	760	427	PEEK, PPTA, PVDF, PES, PEI, PAI, PARA	High-T blends

Notes: *PA-6I, poly(hexamethylene isophthalamide); [†]PPTA, poly(p-phenylene terephtalate) and PMTA, poly (m-phenylene terephthalate); [‡]PBI, poly[2,2′-(m-phenylene)-5,5′ bis-benzimidazol].

Table 2.3. Four categories of amorphous and semi-crystalline polymers

Category	Amorphous	Semi-crystalline
Commodity	PS, PVC, PMMA, ABS, SAN	PP, HDPE, LDPE
Engineering	PC, PPE, COC, (cyclic olefin copolymers)	PEX, PVDC, PBT, PET, POM, PA, UHMWPE
Specialty	PPSU, PEI, PES, PSU	Fluoropolymers, LCP, PARA, PPS, PPA, PA-6/3T
Ultra	PI, SRP, TPI, PAI, HTS, PBI	PFSA, PEEK, PAEK

These CUT limits pertain to neat polymers and may be increased further by compounding with solids and/or blending with another polymer, e.g., PPS has a CUT of 170°C that increases to 260°C in reinforced grades. The reason for the change is that CUT, even in semi-crystalline polymers, is limited by T_g, which after compounding may either increase or become less relevant to some aspects of performance. For example, PEEK with T_g = 143°C and T_m = 340°C has a CUT under load of 150°C. However, PEEK reinforced with carbon fiber (CF) has a CUT ≤ 315°C, at a cost of increased viscosity, density, surface roughness and anisotropy of performance (e.g., weld-lines). An elegant solution is blending, especially if the blend could be made miscible. Good performance has been reported for PEEK blends with another PAEK, as well as with PEI, PI, PSF etc.[3,5–7]

Sauer *et al.* described several miscible blends of PAEK with thermoplastic polyimide (TPI).[18] Thus, PEK, PEEK and PEKK were melt-blended at T > 400°C with three TPIs. The blends had lower viscosities and so were easier to process. More recently miscible HTPB are mentioned below. New hyper-branched polyesters of the AB2 and A2B3 type (the subscripts indicate the number of reactive groups), with T_g = 199–268°C and TGA 10 % weight loss at 400–427°C, were found miscible with semi-aromatic PA6T/6.[19]

The range of HTPBs is shown in Table 2.2. Jaffe *et al.* published a comprehensive overview of HTPB. The authors specifically reviewed PBI and LCP blends.[20] However, as the technology evolves, new polymers and their blends are being developed with better processability and performance. As an example, in 2011 a patent was deposited for blends with new type of PBI (in 1956 the co-inventor, Allan Hay then at GE, patented the polyphenylene ether, PPE).[21] The new polymer, poly(aryletherketone phthalazinone), PAEKP, with T_g = 230–320°C, may be blended with: PAEK, PAK, PEEK, PEKK, or PBI, thermoplastic-PI, PEI, PAES, PPS, and their mixtures thereof. The compositions have improved high temperature processability and performance.

These materials may be molded or extruded into sheets, fibers, foams, tubings, coated articles, etc.

Poly(phthalazinoneethersulfoneketone) (PPESK) is a recent high performance, high temperature resistance polymer, but with high viscosity. However, blends with PEI or PES were found to improve processability and allowed reinforcement with carbon fiber. The blends showed excellent mechanical properties. PPESK was also found to be an excellent basic material for the production of a diversity of membranes with applications ranging from water purification to fuel cell.[22–24]

In 2006 GE announced a new amorphous TPI, Extem®, with $T_g \leq 311°C$, and high strength, stiffness, chemical and creep resistance at $T \leq 230°C$. The new TPI finds applications in auto, aerospace and military products, downhole oil and gas production, medical membranes, electrical connectors, electronics for lead-free soldering, semiconductor wafer handling, and specialty films for insulators and flexible circuitry.[25]

In August 2007 GE Plastics became SABIC Innovative, resulting in further developments in HTP technology. It was found, for example, that Extem® is miscible with PI, PAI, PEI, PEEK or LCP.[26,27] For example, its blends with PEI Ultem® show a smooth increase of T_g from about 220 to 311°C following the Fox equation for miscible blends.[28] These transparent or opaque blends may be custom-colored; they show an enhanced combination of performances at $T \leq 311°C$ including inherent flame-retardancy, wear resistance, electrostatic dissipative properties, dimensional stability, etc. TPI blends with PEEK are available as the Victrex®-MAX series.[29]

Replacing a few bisphenol-A links in PC with resorcinol derived esters leads to isophthaloyl and terephthaloyl resorcinol polyester-carbonate (ITR-PC), also known as Lexan SLX.[30] This polymer is impact, scratch, flame and weather resistant. Furthermore, it is fully miscible with PEI and, having inherent good flame resistance, is suitable for the manufacture of interior components in the aviation industry.[31] Within the full range of composition, $T_g = 142–218°C$, and follows the Fox equation with the correlation coefficient $r = 0.995$. These ITR-PC/PEI blends offered easy processing (melt: $T = 290–370°C$; mold: $T = 80–150°C$), high temperature capability, dimensional stability, high ductility and significantly improved impact properties over PEI.[32] ITR-PC is also miscible with PET, provided that the resorcinol phthalate level ranges from 40 to 80 mol%.[33]

Modification of PC by copolymerization has been highly successful, resulting in six commercial products with others in development.[34] It seems that the strategy was extended to PEI, creating amorphous siloxane-polyetherimide copolymer (Si-PEI), SILTEM® (see Fig. 2.4), with $T_g = 202°C$, 400% elongation at break, flexibility, excellent mechanical performance,

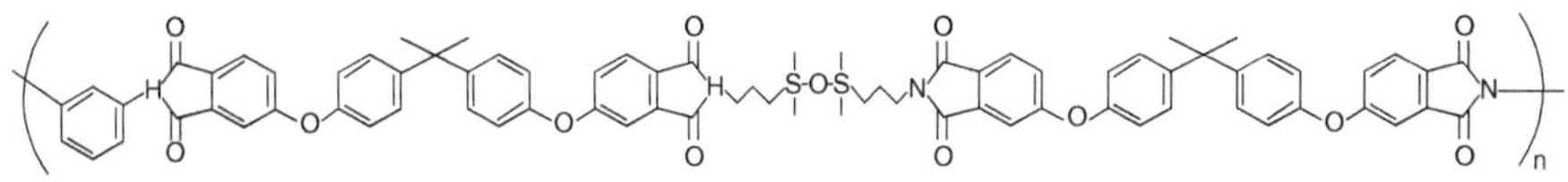

2.4 Amorphous siloxane-polyetherimide copolymer (Si-PEI), SILTEM®.

high thermo-oxidative stability, and inherent flame and chemical resistance. The material was designed for applications in the aerospace, electrical and electronics, healthcare, and automotive industries. The resin was blended with semi-crystalline PEEK (T_g = 143°C; T_m = 344°C), similarly showing good mechanical properties, hydrolytic stability, wear and chemical resistance. The Si-PEI/PEEK blends are fully miscible, their T_g smoothly varies from 143 to approximately 200°C, have excellent properties (namely tensile elongation to break, high softening temperature, flexibility and good processability), are suitable for injection molding, wire coating, extrusion and uses at high temperature, especially where the inherent flame resistance is required.[35,36]

Along with new polymers and their blends there is an important progress in the manufacturing technologies, for example incorporation and orientation of nano-filler particles,[37] gel-spun ultra-high molecular weight polymers (e.g., Dyneema® HB80 cross-plied unidirectional PE fabrics[38], or Innegra S PP fibers with > 90% crystallinity[39]), or control of orientation/crystallization of rigid-rod or platelet macromolecules, etc.

2.3 Methods of polymer characterization

Metrology is the science of measurement. Measurement is based on national or international measurement standards, usually set by national metrological institutes such as National Institute of Standards and Technology (NIST), National Physics Laboratory (NPL), BAM, NPL-I, etc.[40,41] Five independent units of measure are internationally recognized:

- temperature interval (Kelvin, K)
- linear distance (meter, m)
- electrical current (ampere, A)
- time (second, s)
- mass (kilogram, kg).

These SI (Système International) units are fundamental, the others being based on one or more of these. In addition, the SI recognizes candela (cd = luminous intensity) and mole (mol = number of substance elements corresponding to that of 12C atoms in 12 g). Standards are upgraded as scientific knowledge develops; for example, the current definition of meter is the length of the path travelled by light in a vacuum in 1/(299,792,458) s.

The measurement process should comprise:

(1) sampling the test specimens
(2) selection of the standard test method
(3) calibration-traceability of measurements
(4) statistical evaluation of data
(5) validation by means of reference material, inter-laboratory comparisons, etc.[42]

Since 1920 numerous national standardization organizations (e.g., Australia – AS/NZS, Britain – BSI, France – NF, Germany – DIN, Japan – JIS, Korea – KATS, Switzerland – SN, USA – ASTM) have been created, followed by the international standards bodies: European – EN, and International – ISO and IEC. The international organizations try to harmonize measurement procedures where possible.

Polymers are organic materials and their properties, starting with atomic composition up to performance of their multiphase systems, are well documented. However, specific standard test methods are often lacking.[43] Current test methods may be grouped into three categories, with diverse standard test methods available within each of them. These three categories are:

- Determination of molecular composition and structure
- Measurement of material properties
- Materials performance.

In the category 'Determination of molecular composition and structure', standard test methods are:

- Chemical composition by chemical analysis, spectrometry [Analytical spectroscopic methods ISO 6955; Fourier Transform Infrared Analysis (FT-IR) ASTM E1252], chromatography, microanalysis, microscopy, etc.
- Determination of molecular structures on the nano- and micro-scale, using diffractometry, micrography, spectroscopy, scattering and other methods [ASTM 5017 NMR of LLDPE].
- Surface and interphase characterization by spectroscopy (Auger, X-ray photoelectron, secondary ion mass [Surface chemical analysis ISO 17560; Surface Energy ASTM D5946; Surface Resistivity ASTM D257, IEC60093]), and topographic methods [Surface imperfections ISO 8785].[44]

- Processability of polymers [Determination of properties of polymeric materials by means of a capillary rheometer ASTM D 3835–02].

In the category 'Measurement of material properties', standard test methods are:

- Mechanical: elasticity [DIN 16913–2; ISO 6721; ISO 1798 Plastics; Tensile Test Plastics – Microtensile ASTM D1708, Tensile Test Plastics – ASTM D638, ISO 527; Flexural Test ASTM D790 and ISO 178], plasticity [ISO 899 Plastics creep], hardness [ISO 2039 Plastics hardness], strength [ISO 75 Deflection of plastics], fracture mechanics [ISO 13586 Plastics LEFM], impact [Izod Impact (Notched) ASTM D256, ISO 180; Multiaxial Impact (Dynatup) ASTM D3763, ISO 6603, 7765].
- Thermal: conductivity and specific heat [ASTM D2326 cellular plastics], enthalpy, expansion and compressibility [Thermal Expansion ASTM E831, ISO 11359 ASTM D696; D864 Plastics expansion; Compression Set ASTM D395; Compression Test ASTM D695, ISO 604], thermogravimetry [Compositional Analysis by TGA ASTM E1131, ISO 11358], dimensional stability [Dimensional Stability ASTM D1204]
- Electrical: conductivity and resistivity [ISO 21318 for plastics; Volume Resistivity ASTM D257, IEC 60093], dielectric [Dielectric Constant/ Dissipation Factor ASTM D150, IEC 60250; DIN 53483; EN 60811 Insulating materials; Dielectric Strength ASTM D149, IEC 60243], etc.
- Magnetic [EN 62044 Measurement methods].
- Optical: optical sensing, fiber optics, non-linear optics, optical measurements [ISO 3146 Plastics measurements in polarized microscope].

In the category 'Materials performance', standard test methods are:

- Chemical aging, weathering and stabilization [Thermal Stability ASTM D3835, ISO 11443; Artificial weathering ISO 29664; QUV Accelerated Weathering ASTM D4329, D4587, ISO 4892, SAE J2020; Loss of plasticizers ISO 176, ISO 177], oxidation [Thermo-oxidative Stability ASTM D3012, GM9509P], hydrolysis [Water Absorption ASTM D570], accelerated aging [ISO 188 Accelerated Aging], flammability [ASTM D635, ISO 3795, 49CFR-571–302; Oxygen Index ASTM D2863].
- Physical aging [ISO 291:2008 Plastics – Standard atmospheres for conditioning and testing; ISO 877].
- Biogenic effects [Fungus resistance JIS Z 2911PP and PE degradation ASTM D3826; Antibacterial activity ISO 22196; Aerobic degradation of plastics ASTM D 5271, ASTM D 5511, ASTM D 5512, JIS K 6950, JIS K 6951, JIS K 6953].

- Environmental impact [Plastics exposed to solar radiation, heat and light ISO 2578 and ISO 4892, ASTM D 5272].
- Performance monitoring and control [Environmental stress cracking (ESC) ISO 22088; Non-destructive testing by ultrasound DIN EN 12668, ISO 10375].

Evidently, within each of the three test categories there are also standard procedures for testing specific polymers (e.g., PE [ASTM D3035], PP [ISO 1972; ISO 1873] or PVC [ASTM D3036]), or methods for carrying out specific types of test, e.g., Microbeam Analysis (MA):

- ISO 22309:2006 MA – Quantitative Analysis using EDS
- ISO 16700:2004 MA – Scanning Electron Microscopy (SEM) – Guidelines for Calibrating
- ISO 15632:2002 MA – Specification for Energy Dispersive X-ray Spectrometers
- ISO 14595:2003 MA – Electron Probe Microanalysis – Guidelines for the Specification of Certified Reference Materials (CRMs)
- ISO 14594:2003 MA – Electron Probe Microanalysis – Guidelines for the Determination of Experimental Parameters for Wavelength Dispersive Spectroscopy
- ISO/WD2 24173 MA – Guidelines for Electron Backscattered Diffraction Analysis.

Short descriptions of most of the aforementioned standards can be found on the internet.[45]

2.4 Characterization of polymer blends

Polymer blends are mixtures of at least two macromolecular substances. They may belong to the same chemical family (e.g., low and ultrahigh molecular weight PE) or to different ones (e.g., PP/PA-6). The former are known as homologous, the latter heterogeneous polymer blends. At a given set of independent variables, e.g., temperature (T), pressure (P), concentration (x), blends are either miscible or immiscible, i.e., they form one phase or several. The definition of miscibility is based on the thermodynamic free energy of mixing, ΔG_m, and the heat of mixing, ΔH_m.[3,7] Polymers can be categorized as follows.

- The *miscible polymer blend* is homogeneous down to the molecular level, has a negative value of $\Delta G_m \approx \Delta H_m \leq 0$, and a positive second derivative $\partial^2 \Delta G_m / \partial \varphi^2 > 0$. The *immiscible blend has* a positive value of the free energy of mixing: $\Delta G_m \approx \Delta H_m > 0$.

- *Polymer alloys are* immiscible, compatibilized blends with modified interface and morphology.
- *Interphase* is the third phase in binary polymer alloys (Δl = 2 to 60 nm thick), engendered by inter-diffusion of the macromolecular components and/or compatibilization.
- *Compatibilization* modifies the immiscible polymer blend interphase, reducing the interfacial energy and stabilizing the blend morphology.

The above four terms define the key differences between a polymer and PAB. The specific tests developed for the characterization of polymer blends focus on the detection of miscibility, construction of phase diagrams, evaluation of compatibilization and the characterization of interphase.

The commercial nomenclature is not as precise as the scientific one. A polymer is rarely labeled as a blend or an alloy, unless the concentration of the minor component exceeds the locally accepted limiting concentration. For example, 'toughened' PC may be called a blend when the elastomer content exceeds 5 or 45 wt%. Evidently, these products are blends and should be treated as such.[6]

Standard test methods specific to polymer blends do not exist. PABs are tested using the standard procedures developed for polymers or composites. For example, the international standards do not discuss thermodynamic miscibility. There are two standards tests of 'compatibility', a term sometimes substituted for miscibility. However, chemical compatibility tests [ASTM D 543, ISO 4599] refer to evaluation of plastics' resistance to chemicals – tests similar to that for the Environmental Stress Cracking Resistance, ESCR [ASTM D1693, IS22088].[46] Another compatibility test was designed for detecting the effects of other materials (solvents, coatings, solids) on crazing and cracking of PC [EIA 564–1992]. Again, the test has little to do with the determination of thermodynamic miscibility.[47]

All liquids, single or multicomponent, are heterogeneous in terms of the density fluctuations function; thus the miscibility is defined mathematically by $\Delta G_m < 0$, but operationally one likes knowing which size of heterogeneity, d_d. Silberberg and Kuhn proposed that in miscible blends the size of heterogeneity should not exceed that of the radius of gyration: $d_d \leq \langle s^2 \rangle^{1/2} \simeq$ 3–10 nm.[48] The customary scattering methods for detecting miscibility are applicable above that limit, $d_d \geq 15$ nm.[17]

Traditionally, turbidity or opacity has been used for detecting the cloud-point curve (CPC), which approximated the bimodal of the phase diagram.[49] Evolution of this approach involved application of laser light scattering, which combined with small specimen size and precise temperature control led to the pulse-induced critical scattering (PICS) for spinodal determination. Unfortunately, the method is limited to the size of heterogeneity > 100 nm and the difference in refractive index of the two phases > 0.01.[50,51]

The plot of the glass transition temperature, T_g, vs. composition has been frequently used as an unsophisticated test of blend miscibility. However, T_g is insensitive when the amount of the second component is less than about 10 wt% and/or when T_g values of the blend components differ by less than 20°C.[3] Furthermore, it is to be noted that T_g is more sensitive to the degree of dispersion than to the thermodynamic miscibility of the components.[52]

Recently the Atomic Force Microscopy (AFM) showed that in a single-component molten PS below the crossover temperature, T_c, there are solid dynamic aggregates with diameter $d \approx 4$–40 nm.[53] Since on the one hand blend miscibility has been postulated on the basis of a single T_g, and on the other hand the presence of a single T_g has been detected when the degree of dispersion $d \leq 15$ nm, the presence of such aggregates in neat PS blends makes detection of miscibility by the T_g method ambiguous. A better measure of blend homogeneity is the width of T_g peak at half-height, ΔT_g. Its value in homopolymers, miscible blends and immiscible blends near the spinodal curve are, respectively: $\Delta T_g = 6, 10$ and 32°C.

The advanced scattering methods, such as high resolution nuclear magnetic resonance techniques (HR-NMR), the NMR-spin diffusion, non-radiative energy transfer, excimer fluorescence, thermally stimulated depolarization current, small angle neutron scattering, SANS and FT-IR, are more appropriate for the task. For example, the NMR spin-lattice relaxation times, T_1, distinguishes $d_d \geq 2$ nm and it may be used for either molten or solidified specimens:[54]

- a-PVC/PMMA: homogeneous at 20 nm, but heterogeneous at 2 nm, but a-PVC blended with deuterated d-PMMA was miscible, $d_d \approx 1$–2 nm;
- PS/PVME, poly butyl methacrylate (PnBMA)/PS (2 mol%-OH): homogeneous at 20 nm;
- SAN/PMMA: homogeneous at 20 nm, but heterogeneous at 2–15 nm;
- a-PET/PC, a-PVDF/PMMA, and a-PCL/PVC: homogeneous at 2 nm.

In addition to the direct experimental methods of miscibility detection, there are several fundamental approaches useful for predicting or justifying miscibility/non-miscibility of a given blend. For example, ΔH_m was used by Paul and Barlow for predicting blend miscibility:[55]

$$\Delta G_m \approx \Delta H_m \leq 0 \text{ when } 2B_{11} < B_{12} + B_{22}. \qquad [2.1]$$

The Huggins–Flory type segmental, binary interaction parameters, Bij, are tabulated.[56] A similar approach, based on the Hildebrand solubility parameters (∂), calculates ΔH_m from the high pressure dilatometry (*PVT*) data.[57]

Atomistic calculations of the interaction energies and enthalpies of mixing for PAB modeled by realistic pair configurations of two polymer fragments may predict the specific interactions from tabulated force field constants.[58–60] The diverse atomistic simulations of blend energies of mixing are powerful tools for predicting a host of diverse properties, including miscibility.[61] The simulation results depend on the selected procedure and completeness of the energetic contribution. For example, the recent simulation of a 1:1 blend of poly-N-vinyl-pyrrolidone (PVP) and poly(bisphenol-A-ether-sulfone) (PES) with a COMPASS (condensed-phase optimized molecular potentials for the atomistic simulation studies) force field expressed the total energy of the system as a sum of seven bonding and non-bonding interactions.[62,63]

The statistical mechanical Polymer Reference Interaction Site Model (PRISM) also predicts structure and phase diagrams of polymer blends.[64,65] The theory involves computations of the system structure by means of the probability density function, which describes the location of all N particles of the system. The computations include enthalpy, as well as the combinatorial and non-combinatorial entropy effects. Over the years RISM has been applied to progressively more complex liquids, from mono-atomic to multi-atomic molecules, then to polymers and their blends. PRISM requires less computing time than atomistic simulations, incorporating structural details of polymeric molecules (e.g., bond lengths, angles, chain conformation, different interaction potentials), which lattice models are unable to account for PRISM results of polymer blends are well correlated with the small angle neutron scattering (SANS) and light scattering (LSc) experimental data.[56]

2.5 Characterization of HTPBs: chemical constitutions and molecular weights

Table 2.4 provides an abbreviated list of methods used for characterization of polymer blends, starting with determination of molecular structure and interactions and ending with determination of the crystallinity and flow-imposed orientation. The methods are general – their adaptation to HTPBs may require modification of equipment (e.g., high temperature cell for FT-IR or special heating oven for rheology of PEEK).

There are two aspects in the characterization of polymers and their blends: (1) chemical constitution, (2) molecular weight (MW) and its distribution (MWD). Identification of the chemical constitution is conducted by the manufacturer and rarely performed by the industrial or academic users.

Table 2.4 Characterization of polymer blends

Method	Molecular structure interactions	Miscibility	Surface and interphase		Phase separation	Crystal structure	Orientation
Scattering	SANS	E	E	G	G	E	E
	SAXS	E	E	G	E		
	SALS		G	G	F		
Microscopy	TEM	G	G		E	E	E
	SEM	F	G		G	G	G
	PCM	F	F		F	F	F
	PM	F	F			E	G
Relaxation	NMR-pulsed	G	E	G	G	E	F
	Rheology	F	F	F	G	G	G
	Dynamic mech.		G		G	G	F
	Dielectric spect.		G				
Thermal analysis	DSC, DTA	F	E	F	G	G	G
	TOA		G	F	F	G	
	TSC		G		G	G	
Spectroscopy	FT-IR	E	F		G	G	G
	HR-NMR	E	E		F	G	G
	ESCA	G	E		E		
	SIMS		E		E		
	Fluorescence				F		

Notes: E = excellent; G = good; F = fair; SALS, small angle light scattering; PCM, polarized contrast microscopy; TOA, thermal oxidative analysis. Other abbreviations are listed in Section 2.12.

Source: T. Nishi, *J. Appl. Polym.* Sci., **79**, 192 (1993).

The exception may be the determination of fine isomeric sequences (e.g., in polyolefins, PO), identification of tactic content (for example in PVC), or identification of end groups (e.g., in PA, PEST, or PO). The applied analytic methods include FT-IR, SANS and pulsed nuclear magnetic resonance (NMR). These may be combined with pyrolysis, followed by chromatography and mass spectrometry (MS). For the users of HTP, the most pertinent information is how the polymeric structure, MW and MWD, changes with time.

Quantitative NMR (e.g., 1H NMR and 13C NMR spectroscopy) has been used for determining monomer/co-monomer composition, polymeric structures and impurities down to 0.1–0.01%. For enhanced resolution of 13C spectra inverse-gated 1H decoupling may be used. A 500 MHz NMR instrument at $T = -80$ to +150°C is suitable for analyzing polymers in solution, e.g., PE or PP dissolved in 1,2,4-trichlorobenzene and analyzed at 130°C. ASTM D5017–2009 recommends using 13C NMR spectroscopy for determining impurities or co-monomer ratios of linear low density PE (LLDPE).

Historically, the following methods have been used for measuring MW and MWD:

- size exclusion (or gel permeation) chromatography, SEC or GPC;
- light scattering (LSc);
- viscometry (intrinsic viscosity: $[\eta]$);
- mass spectroscopy (including the matrix assisted laser desorption, MALDI);
- end group analysis (chemical and spectroscopic methods);
- colligative properties (boiling point elevation, depression of freezing point, osmometry);
- others.

As will be discussed in Chapter 3, HTPs are often employed in severe environments; thus their structure, MW and MWD, may alter their performance. The polymers and their blends may degrade by either reducing the average MW, or increasing it by branching or crosslinking. For example, PS has been known for over a century for its ability to depolymerize into styrene, whereas most elastomers are known for crosslinking. These effects may be followed by the dynamic melt flow, or viscoelasticity measurements. The decrease of loss modulus (dynamic viscosity) suggests depolymerization or chain scission, whereas increase of the storage modulus (dynamic elasticity) indicates branching, which eventually leads to crosslinking. Thus, the dynamic rheology may be used as an indirect measure of MW and MWD.

During the last two decades, there was a concerted effort for predicting flow behavior from MW and MWD and vice-versa, calculating the two molecular parameters from rheology.[66–70] In that work, invariably the parameters calculated from rheology were compared with those obtained directly from SEC (previously known as GPC).[71] SEC (introduced to polymers by Claesson and Claesson in 1944) belongs to liquid chromatography methods, where an injected solution of polydispersed polymers is separated by macromolecules being entrapped in pores of porous gel particles (e.g., crosslinked PS) – the smaller the molecule the deeper it penetrates and the longer it stays in the column. SEC in combination with low-angle laser light scattering (LALLS) and/or solution viscometry determines the MW and MWD as well as the degree of long chain branching of a polymer.[72–75]

The development of the solution viscosity theory[76] relates the coil volume to intrinsic viscosity:[77–79]

$$[\eta] = KM_v^a , \qquad [2.2]$$

opening the possibility of determining the viscosity-average MW (M_v) of the dissolved polymer. Its fractionation into narrow-MWD fractions, followed by determination of $[\eta]$ of each, resulted in the early determination of MWD curves.[80–82] In Equation [2.2], $[\eta]$ is the intrinsic viscosity and K and a are constants, whose values depend on the nature of the polymer, solvent and temperature.

An inverted variant of SEC, the temperature rising elution fractionation (TREF), has been developed for determining the short chain branches population distributed over the various molecular weights.[83] Polymer heterogeneity is not only represented by the distribution of chain lengths (MWD), but also by tacticity, presence of branched and cross-linked structures and, in the case of copolymers, by differences in chemical composition. Thus TREF was developed for examining how the diverse heterogeneities superpose on the MWD. It is a technique for analysis of crystalline and semi-crystalline polymers by separating the fractions according to crystallizability.[84] The molecular elution during the test depends on the melting of semi-crystalline polymers in the presence of a solvent. The customary detectors are infrared, light scattering and a capillary viscometer. Thus, by contrast with SEC, in TREF fractionation takes place by melting and dissolution. Newer versions of TREF are preparative (PTREF) and analytical (ATREF), the latter applicable for analyzing blends of crystallizable polymers.[85]

Several types of extractors and test strategies are used for characterization or purification of a variety of polymeric materials, including copolymers, blends, nano-composites, etc. For example, the Kumagawa extractors have been used for elimination of the low molecular weight fractions.[86] The special thimble holder is surrounded by solvent vapors at a higher temperature than that in Soxhlet; thus this extractor is better suited for removal of low molecular weight impurities from HTP.[87] Vandenburg *et al.* published a critical review of old and new extracting methods, including precipitation from solution, the use of Soxhlet-type extractors, the supercritical fluid extraction (SFE), microwave-assisted extraction (MAE), accelerated solvent extraction (ASE), sonication-assisted extraction, etc.[88] The methods vary in the extraction time (from 15 min up to 24 h), range of application and the development stage. The extracted substances are: monomers, dimers, trimers, plasticizers, industrial additives, stabilizers, etc.

The inverse gas chromatography (IGC) also involves deposition of polymers or their blends on solid support and probing with suitable solvent vapors. The method, first described by Kiselev in 1967, has been used in blend technology for characterizing the polymer-polymer and another type binary interactions.[89–93]

Reed and coworkers developed a strategy for automatic continuous online monitoring of polymerization (ACOMP).[94,95] The method may be used during the initial development of the polymerization process, its optimization and monitoring of the continuous reaction. ACOMP automatically dilutes samples from the reactor and measures its properties, e.g., refractive index, near infrared (NIR) spectra, LSc, $[\eta]$, etc., from which it computes evolution of MW, MWD, degree of conversion, copolymer composition (in copolymerization) and others. The method has been applied to a variety of the free radical homo- and co-polymerizations, including the reactions in emulsion or suspension.[96–98]

2.6 Characterization of HTPBs: chemical-, thermal-, mechanical- and radiation-induced degradation

Polymer' degradation may be categorized according to the degrading agent as:

- chemical (e.g., oxidation or hydrolysis);
- thermal;
- mechanical (e.g., by ultrasonics or during processing);
- radiation (ultraviolet, visible light, nuclear or cosmic rays, atomic oxygen, etc.).

Since HTPs are often used under severe conditions (temperature, *T*, pressure, *P*, irradiation, *Q*, stress, σ, and their combinations), their structure, MW and MWD, change with the lifetime. The process may lead to the reduction or increase of MW, erosion, etc. Several mechanisms have been identified for either effect, and stabilization methods are constantly upgraded.[99,100]

2.6.1 Chemical degradation

The chemical degradation takes place when a polymer is exposed to reacting fluids, such as gaseous O_2, O_3, Cl_2, H_2S, H_2O, etc. In polymers with carbon-carbon backbone, oxidation is the principal chemical degradative mechanism that results in formation of peroxides and/or hydroxy peroxides, chain scission and accompanying it the free radical grafting (which may be absent in high vacuum). Polymers with tertiary carbons (e.g., PP or linear low density PE, LLDPE) oxidize readily; for PO the process is autocatalytic.[101]

It is noteworthy that O_2 dissolves in polymers to a rather large depth, namely 10 mm in PP.[102] Repeated re-extrusion of polystyrene (PS) in a twin-screw extruder (TSE, under nitrogen) each time reduced its MW by around 10%. The reduction was not related to the level of stresses in TSE and, since in a rheometer the dynamic properties were constant with time, the observed effects could only be explained by the presence of absorbed oxygen, O_2, in the pelletized polymeric compounds – once the resident oxygen was consumed, the compound became stable.[103] Thus, oxygen dissolution in polymers is a common phenomenon, expected in amorphous or meso-crystalline regions. Polymers with double-bonds readily suffer degradative oxidation and ozonolysis, crosslinking by sulfur, etc. Oxidation is also present in PA, PEST or poly-glycols such as polyethylene glycol (PEG), but in these cases the hydrolysis dominates.

Hydrolysis may be treated as de-polymerization, in which water molecules peel off molecular fragments, e.g., for polyester: $R_1 CO_2R_2 + H_2O \rightleftharpoons R_1CO_2H + R_2OH$. However, since the reaction is a reversible one, condensation and hydrolysis take place at the same time and the equilibrium is determined by the concentration of products on both sides of the equation. In polymer processing, hydrolysis of the condensation polymers, e.g., PA, PC, PEST, occurs during high temperature processing, namely extrusion or injection molding.[104] To prevent their loss of performance these polymers must be dried before processing to the maximum acceptable level of moisture for the given polymer, e.g., 0.2 wt% or 2000 parts per million (ppm) for PC, 1000 ppm for PA-6 and PA-66, 400 ppm for PBT, 30 ppm for PET, etc.

2.6.2 Thermal degradation

A good introduction to this topic is provided by Beyler and Hirschler, with a short introduction to polymers and their properties.[105] Thermal degradation/stability has been studied by several methods, namely TGA, differential thermogravimetry (DTG), thermal volatilization analysis (TVA), differential thermal analysis (DTA), differential scanning calorimetry (DSC), pyrolysis gas chromatography (PGC), thermo-mechanical analysis (TMA), and others. Furthermore, the positron annihilation lifetime spectroscopy (PALS) was found useful for the thermal degradation studies of PE. PE films exposed to $T = 100°C$ for $t \leq 30$ days showed significant reduction of the *o*-positronium formation caused by the thermal oxidation of PE. The method is sensitive to degradation of PE and other nonpolar polymers and their blends.[106]

McNeill and Rincon studied thermal degradation of PC by means of TGA, TVA, DSC, FT-IR, mass spectrometry (MS) and gas chromatography–MS (GC-MS) method (see Fig. 2.5). PC is stable up to 300°C.[107] Above that temperature, small quantities of phenol and *p*-cresol were detected, at 375–400°C CO_2 appeared and then at $T \geq 455°C$ CO and CH_4 were formed; the peak on the DTG curve was at $T = 462°C$. At 500°C the main products are cyclic dimer and bisphenol A, with small quantities of CO_2, *p*-cresol, *p*-ethyl phenol, phenol, *p*-vinyl phenol, *p*-isopropyl phenol, CO and CH_4. In the absence of air and moisture, degradation of PC proceeds by hemolytic decomposition of the polymer chain, radical reactions, fragmentations and molecular rearrangements.

However, PC degradation is also sensitive to the presence of oxygen and water. Accordingly, there are patents on stabilization of PC against thermal as well as hydrolytic degradation. The thermally stable PC composition might contain a dioxane phosphate (prepared in reaction between tri-phenyl phosphate with a 1,3-dioxane and phenol),[108] or a sulfone compound (of the general formula: R_1-SO_2-CHR_2R_3, including polymers with sulfonyl groups).[109] Both these stabilizers belong to the peroxide decomposer category. The patent for hydrolytically stable PCs specifies addition of a stabilizing compound selected from epoxy silanes, epoxy siloxanes or their blend.[110] The stabilizer of a general formula: Z_nSiR_{4-n}, (where Z is epoxy radical and R is hydrogen, alkyl or aryl radical) is added in the amount of 0.005 to 2 wt%. For good color stability the PC formulation should also contain an organophosphite.

The most frequently used is TGA at constant temperature or heating rate of $q = 5$–$20°C/min$ in vacuum, air or inert gas.[111] The degradation process has been described taking into account the reactions: (i) initiation by random chain scission, (ii) depropagation, and (iii) chain termination.[112] TGA is an integral part of synthesis and characterization of HTP and their blends. For example, developing new isomeric polythioetherimides (PTEI),[113] polyamides based on *s*-triazine,[114] polyaryletherketones (PAEK)[115] or polyimides (PI)[116,117] and their

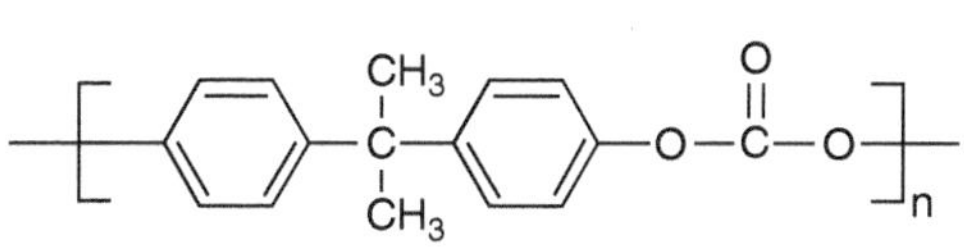

2.5 Polycarbonate of bis-phenol-A (PC).

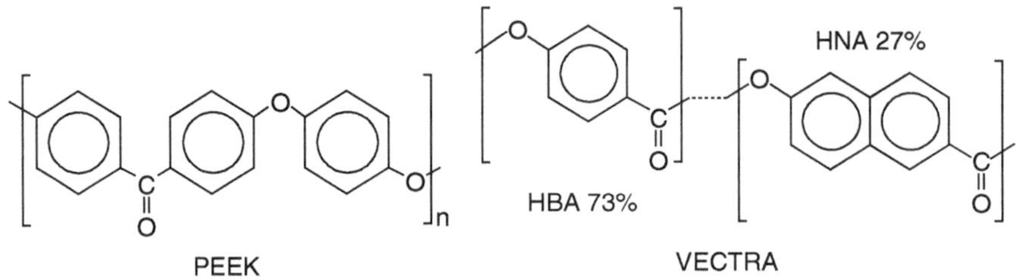

2.6 Molecular structure of poly(ether-ether-ketone) and Vectra LCP.

nanocomposites[118] invariably involved determination of thermal stability by TGA or its differential version. Naffakh *et al.*[119] studied the thermal decomposition of PEEK blends with a thermotropic Vectra® LCP, using dynamic TGA and calculating the activation energies of the process. The molecular structure of the polymeric constituents is shown in Fig. 2.6. Compared to neat components the blends were found to be less stable. In PEEK/LCP = 85/15 wt/wt degradation of the former high melting point component was accelerated by the latter. The FT-IR indicated that in blends the degradation rate was higher, but the mechanisms remained the same.

While PC is stable up to 300°C many polymers that are blended with it decompose at lower temperatures.[120] For example, the miscible blends of PC with PMMA were studied by TGA, TVA and DSC. During thermal degradation of the blend the volatiles were the same as those from component degradations, the rate of PC degradation increased while the PMMA depolymerization was retarded by around 10°C. The results suggest that PMMA radicals accelerate chain scission of PC while PC retards PMMA degradation. The PC/PMMA blend is unique for the reaction between a macro-radical and a macromolecule, which are seldom observed during degradations of immiscible blends.

PC being polyester may also undergo trans-reaction with another polyester, polyester-amide or polyamide. Lin and Yee studied such reactions in blends of a liquid crystalline copolyesteramide (LCP, Vectra® B950) with PC or PET.[121] Due to transesterification, the PC/LCP blends were slightly miscible, with miscibility increasing with time and temperature, $T < 340$°C. The trans-reaction also reduced LCP crystallinity. The DSC-determined decomposition temperature of LCP/PET blend was about 400°C. The exchange reactions between LCP and PC or PET improves the interfacial adhesion in these immiscible blends; thus they may be used for *in situ* formation of compatibilizing copolymers. The reactions are rather slow and need to be accelerated by, for example, an ionic catalyst.

Evidently, all reactants in polymer blends contribute to thermal degradation/stability, but it is not *a priori* evident how they will affect each other. For example, PVC stability was poorer in its blend with polyvinyl acetate

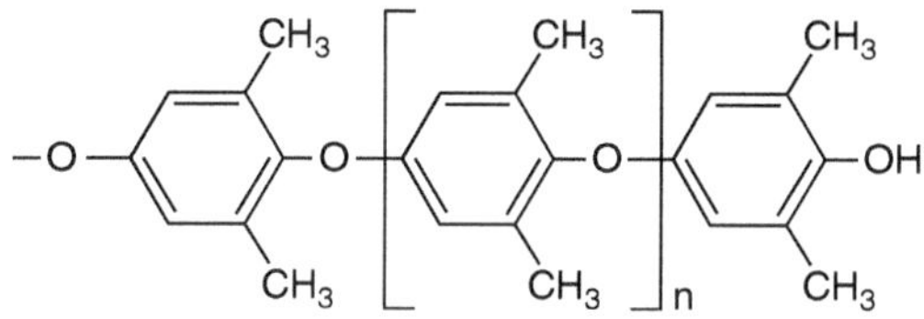

2.7 Poly(2,6-dimethyl-1,4-phenylene oxide (PPE).

(PVAc) than without it.[122] TGA of PP, PE, polyvinylidenechloride (PVDC), and polyacrylonitrile (PAN), as well as of their blends, were carried out. Comparing with neat PP its blends with PE had higher thermal stability, in contrast to the results for PVDC with PAN.[123] Syndiotactic polystyrenes (sPS) were solution blended with PPE (Fig. 2.7), co-precipitated, and then compression molded. DSC indicated a partial crystallization of sPS and some miscibility in the amorphous phase.[124] TGA (at q = 10°C/min) and FT-IR spectroscopy were used for following the structural changes on heating. Degradation of the three molecular weight sPS resins was similarly initiated by chain scission. In the blends, degradation of sPS started at higher temperature by 10–20°C, indicating stabilizing effects of PPE, but in turn degraded faster in the blends than in the neat state. FT-IR identified the structural changes of PPE molecules at T = 250–400°C as breakage of the ether link followed by H-abstraction from the methyl, which results in an increase of the –OH group concentration. Since sPS degradation and PPE rearrangement occurs at the same T, it seems that the latter polymer is a radical acceptor, slowing the degradation of sPS. This cross-termination process limits the PPE rearrangement and consequently reduces the ultimate stability of that polymer.

Bacosca *et al.* synthesized a series of aromatic PI with cyano-groups, by polycondensating aromatic di-amine having cyano and ether groups, with either flexible or rigid aromatic di-anhydride.[125] The flexible polymers were soluble in polar organic solvents and easily formed into films with high thermal stability. Their T_g = 210–260°C and the initial decomposition temperature, T_{decomp} > 400°C, increased with the increase of the chain rigidity. The rigid PI, based on pyromellitic *di*-anhydride (PMDA, see Fig. 2.8) had T_g > 250°C and it was insoluble.

2.6.3 Mechanical and thermomechanical degradation

It is no surprise that a combination of different agents that cause polymer instability increases the degradation rate. For example, the presence of heavy metals (left behind catalyst or brought in with, for example, clay nano-particles) enhances photo- or thermo-oxidative degradation.[126] Dynamic flow behavior of the polymeric nanocomposites (PNC) with either PS or PP matrix

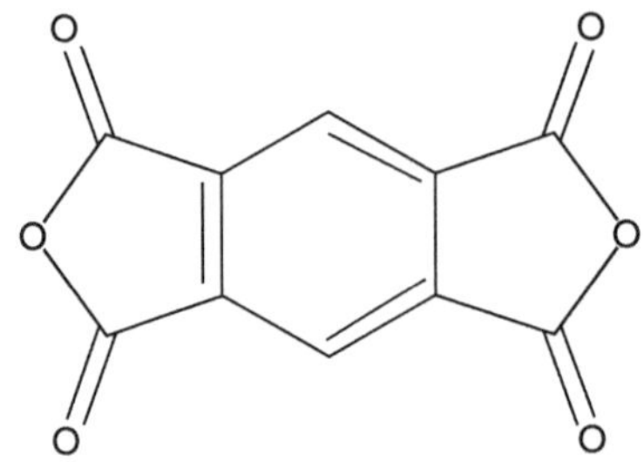

2.8 Pyromellitic *di*-anhydride (PMDA).

showed that during time sweep in a rheometer (under a blanket of N_2) the storage and loss shear moduli (G' and G'', respectively) were constant, indicating thermal stability. However, each re-extrusion in a TSE under N_2 reduced MW of the matrix polymer by around 10%. The reduction was not related to screw configuration or speed. These seemingly conflicting results could only be explained by the presence of absorbed oxygen, O_2, in the pelletized polymeric compounds – once the resident oxygen was consumed, the compound became stable.[103,127,128] Data analysis showed that the principal degrading agent was absorbed O_2 activated by shear and temperature.

When studying the effects of the mechanical stress on polymer degradation, it is difficult to separate the mechanical effects from the thermal effects.[129] So most publications on the subject discuss these two factors together as the thermomechanical process. Several key publications on the thermomechanical behavior of PAB came from the Università di Palermo. For example, in high density PE (HDPE) blends with PP the chain scission of PP molecules dominated composition with more than 20 wt% PP, whereas below that limit branching of HDPE overpowered these effects. [130] Figures 2.9 and 2.10 compare the effect of composition on the degradation process. As is evident, blending of two POs, PE and PP, at constant T and screw speed indicate a mild, if any, binary influence on degradation, but the situation is dramatically different in blends of PA-6 with PC. In this case, the presence of around 35 wt% PA-6 causes maximum degradation of PC. The effect may be caused by transreaction, catalyzed by the presence of H_2O.

La Mantia and Mongiovi studied the thermomechanical degradation of PA-6/PP blends without and with a compatibilizer. The degradation depended on the compatibilizing functional group. Thus, degradation was worse in the presence of acrylic acid (AA) then maleic anhydride (MAH) groups on PP. In other words, compatibilization and degradability inseparably influenced each other. In the case of PA-6/PP-*g*-AA blend the compatibilization (and degradation) decreased with mixing time, caused by the shear effects on the PP-PA-6 copolymer. Blends with PP-*g*-MAH showed relatively good resistance to thermomechanical degradation and better performance than those

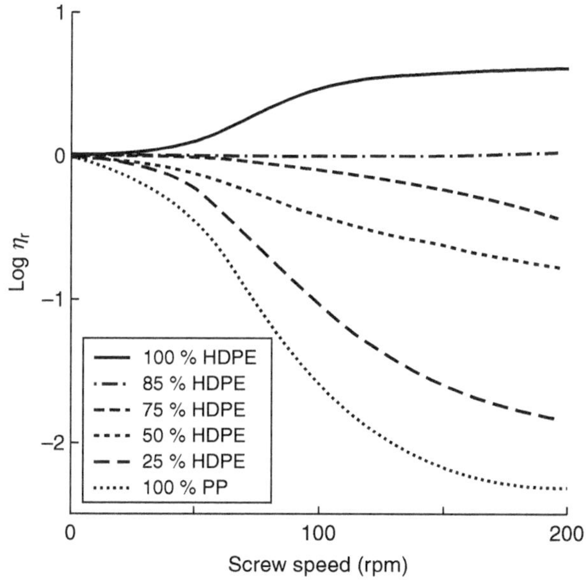

2.9 Effect of composition and screw speed on the relative viscosity of HDPE/PP blends at 280°C. After La Mantia and Valenza.[130]

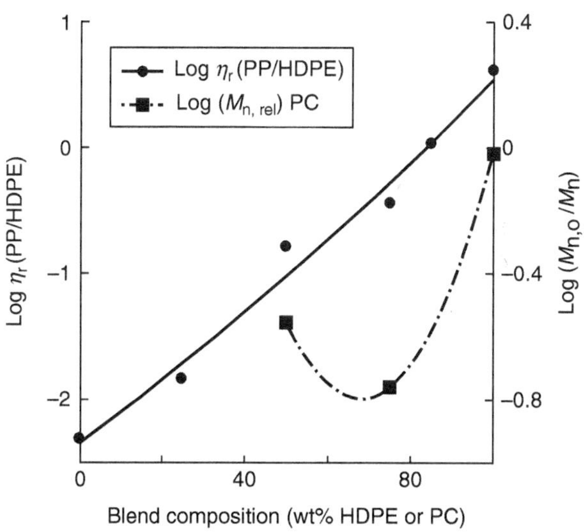

2.10 Effect of composition on relative viscosity of HDPE/PP blends and the number-average molecular weight of PA-6 blends with PC. Adopted from La Mantia and Valenza.[130]

with PP-*g*-AA. The binary PA-6/PP blends initially showed improved properties, which decreased with mixing time, which reduced MW.[131]

2.6.4 Degradation by radiation

Polymers, their blends and composites may be exposed to radiation during radiation processing, or when exposed to irradiation (e.g., irradiative sterilization, exposure to radiation in space, exposure to radiation during wire coating).[132] In space (hard vacuum and cold) the radiation may lead to mass loss (progressive surface erosion). However, in inert atmosphere, irradiation may cause crosslinking, degradation or both.

- *Crosslinking*: PA, PE, PEG, PEST, PP, PS, polyacrylates, polyacrylamides, polysiloxanes, rubbers, and so on.
- *Degradation*: polyisobutylene (PIB), poly(α-methyl styrene), polymethacrylates, polymethacrylamide, polytetrafluoroethylene (PTFE), and so on.

Since irradiation leads to chain scission and formation of free radicals, in the presence of O_2, it leads to the formation of peroxides, which further accelerate the degradative effects.

The radiation affects the miscible and immiscible polymer blends differently.[133,134] Whereas in the former case the effect is similar to that encountered by individual homopolymers, in the latter it may be used for improving compatibility, and thereby the performance of blends or for stabilization of their morphology by a radiation crosslinking. PP is particularly sensitive to irradiation, which makes it discolored and brittle. However, the behavior may be improved by blending PP with polybutylene (PB), then quenching into tough and radiation-resistant non-crystalline alloy.[135]

A large volume of literature exists for PO blends with easily crosslinkable elastomers, but the information on radiation processing and degradation of HTPB is scarce. For dimensionally recoverable applications a blend of LCP with PVDF or PE was radiation crosslinked.[136] Similarly, PSF or PES was blended with water-soluble PVP and crosslinked to make it water insoluble for medical or food applications.[137] For the use in cable jackets and heat-shrinkable applications PVDF or a copolymer of tetrafluoroethylene and ethylene (ETFE) could be compounded with a thermoplastic elastomer, formed and radiation crosslinked. Adjusting composition and irradiation dose produced a series of materials with good balance of tensile strength and elongation.[138]

Sterilization of PC by irradiation causes yellowing and loss of performance. Analysis of gases generated during irradiation showed the presence of gases: $CO > CO_2 >> H_2 >> CH_4 >$ benzene, as well as phenolic and salicylic esters coupling by-products. The process engenders

Fries-type free radical reactions that affect the sensitive carboxylic groups.[139] However, the UV stability of PC may be significantly improved if during its synthesis aliphatic and aromatic ester glycols are incorporated into the PC chains.[140] Another patent on a production of γ-radiation-resistant PC disclosed incorporating of disulfide linkages (0.02 to 2%) into the main chain, side or terminal groups.[141] The presence of the disulfide bisphenol improved the resistance to discoloration caused by γ-rays. The resulting PC-disulfide resin may be used directly for molding radiation-resistant articles or as a stabilizing additive for neat PC. Another patent described incorporation into PC macromolecules 0.01 to 1.0 % of benzofuran and 0.01 to 1.0 % of saccharine.[142] This modified PC has good resistance to degradation by γ-radiation as evidenced by reduced yellowing. The new material is suitable for medical devices.

PC exposed to neutron irradiations with the incident energy in the range of 0.8–19.2 MeV degrades by the free radical processes that lead to yellowing, crosslinking and embrittlement. The process was followed by FT-IR, X-ray diffraction (XRD), solution viscosity, refractive index and color change studies. The FT-IR showed that in the energy range of 2.3–19.2 MeV crosslinking dominates increasing $[\eta]$ (from 0.41 to 0.68) and refractive index.[143] The yellowing could be reduced by blending PC or its copolymers with poly(cyclohexane terephthalate-glycol), PCTG. These well-processable, miscible blends (e.g., *Ektar® DA* for garden, floor care, appliances, medical equipment) are transparent, resistant to UV, neutron and γ degradation and have good chemical resistance.[144,145]

The Industrial Applications and Chemistry Section of the International Atomic Energy Agency published a series of articles on stability and stabilization of polymers and their blends and composites.[146,147] Most of the articles there discuss degradation mechanisms and stabilization methods for commodity polymers and their blends. The exception is the work by Sun and Zhong on γ-irradiation crosslinking of fluoro-polyimide (FPI).[148] The authors reported that the reaction resulted in an increase of T_g, the tensile strength, the elongation at break, resistance to hot water as well as to subsequent irradiation. The X-ray photoelectron spectroscopy (XPS, also known as ESCA) showed that during radiation crosslinking the concentration of the terminal -CF3 groups decreases, suggesting defluorination. Similarly, PTFE may be radiation-crosslinked at high temperature for better performance and radiation resistance. It is noteworthy that irradiation of the same polymer may result in chain scission or crosslinking – the key is the temperature:

- at $T < T_g$ irradiation reduces MW and reduces performance;
- at $T > T_g$ branching and crosslinking increases MW and improves performance.

The effect is related to chain flexibility. Accordingly, miscible blends of rigid with flexible backbone polymers improve the radiation resistance, namely, blending two fluoro-co-polymers or PMMA with PEG or with PVDF yields radiation-resistant blends.

Alvino studied the stability of two PIs and PAI exposed to ultraviolet (UV) irradiation in wet or dry environment.[149] The films were UV-irradiated (290–400 nm) for 125 and 250 days, respectively. The PIs were found to be more stable than PAI, but the latter was less susceptible to hydrolysis. The three polymers were more sensitive to UV in the wet than in a dry state. The degradation affected the elongation at break more than the tensile strength. By contrast, the electrical properties of dry PIs were unaffected by UV exposure at least up to 250 days while PAI begin to deteriorate after 167 days of exposure. The structure of a PI and a PAI are shown in Figs 2.11 and 2.12.

Hill *et al.* simulated low Earth orbit (LEO) radiation effects on PI films of different molecular structures, studying UV degradation in air and under vacuum.[150] The process was followed using UV-visible, FT-IR, electron spin resonance (ESR), and XPS spectroscopies. The UV-visible spectrum showed a blue shift relative to Kapton, caused by the structural disruption of a charge-transfer complex. The UV irradiation also resulted in an ablative surface degradation with intact central part of the PI films. Irradiation in vacuum caused formation of stable carbon-centered radicals, while that in air produced oxygen-centered radicals of peroxy and/or alkoxy type; XPS indicated the surface oxidation accompanied by aromatic rings opening and loss of low molecular weight oxidized carbon species.

2.7 Stabilization of HTPBs

Stabilizers are chemicals used for maintaining the polymer's strength, flexibility and toughness during their processing and life-time; thus they should preserve the polymer's original molecular structure. In other words, stabilization aims to retain the original molecular architecture in spite of the effect of heat, stress, chemical environment and irradiation. Stabilizers are classified as:[151]

- primary antioxidants (AO1) – phenolic or 2,6-di-tert-butyl-4-methylphenol (BHT) type;

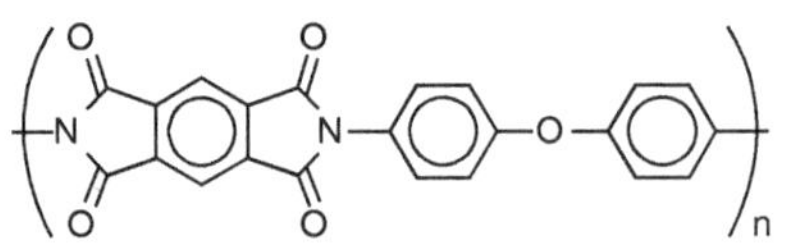

2.11 DuPont polyimide, PI, Kapton.

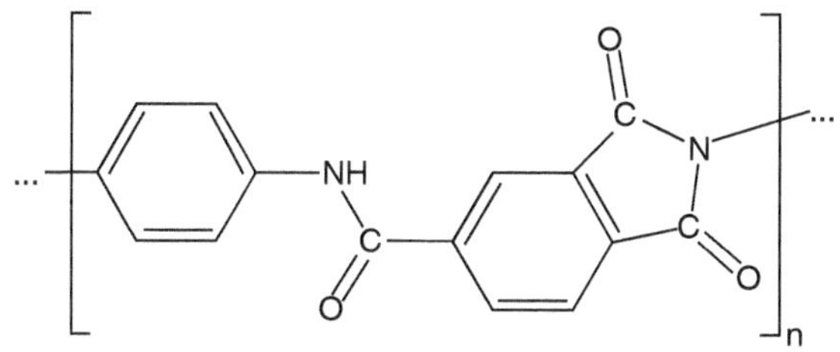

2.12 Solvay polyamide-imides, PAI, Torlon.

- secondary antioxidants (AO2) – phosphites, phosphonites, sulfur compounds;
- metal deactivators or complexing agents;
- Ni-quenchers or hydroxyperoxide deactivators;
- UV absorbers (UVA) – transparent absorbing UV within 300–400 nm range;
- hindered amine (light) stabilizers (HALS, now HAS) – radical scavengers.

Without a suitable stabilization package, polymers exposed to light and O_2 would degrade. Appropriate mixtures of stabilizers protect polymers at various stages throughout the processing and formation stage as well as during their lifecycle. For example, an organophosphite is a short-term antioxidant that protects the polymer during the high temperature and shear processing, while phenolic antioxidants offer a long-term defense.

The stabilizing package must be designed to finely disperse the appropriate stabilizer at a specific location or phase where it is needed. For this reason, its melting point (T_m) should be below the processing/forming temperature and it must be thermodynamically miscible with the degradation-sensitive polymer. Since the stabilizing activities are controlled by functional groups, adjustment of miscibility depends on the type and size of paraffinic vs. aromatic parts of the stabilizer molecules. Table 2.5 gives examples of four popular stabilizer types. The listed boiling point (T_b) indicates the upper temperature that would eliminate the stabilizer from the compound; thus it should not be approached either during processing or application.

Degradation of polymers starts with a formation of free radicals, $R^{\bullet}$, caused by structural defects, temperature, stress, etc. Examples of labile bonds that break creating $R^{\bullet}$ are listed in Table 2.6. Once $R^{\bullet}$ is formed, the radical degradation process is initiated; it abstracts the most labile H-atom in the stabilizer or polymer. In the presence of O_2, $R^{\bullet}$ transforms into peroxy radical, $ROO^{\bullet}$, which accelerates the autocatalytic process. The alkyl-peroxy radicals dominate the propagation step of degradation. Usually polymers

Table 2.5 Examples of plastics stabilizers from CIBA *Plastics Additives Handbook*[151]

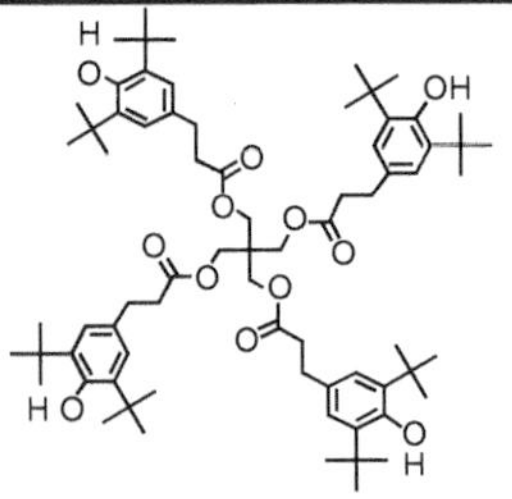	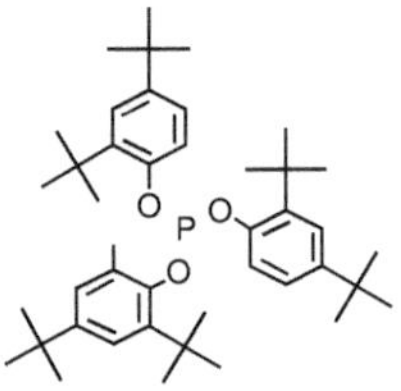
Irganox 1010, tetraphenol antioxidant CAS#: 98584-37-3; $(C_{17}H_{25}O_3)_4C_5H_8$ MW: 1177.6 [g/mol]; T_m = 110–118°C, T_b = 1130.4°C (calculated)	**Irgafos 168** = tris(2,4-*di-t*-butylphenyl)phosphite CAS#: 31570-04-4; $(C_{14}H_{21}O)_3P$ MW: 646.9; T_m = 181–184°C, T_b = 619.8°C (calculated)
Recommended stabilizer: **Irganox B 225** = 50% Irganox 1010 + 50% Irgafos 168	
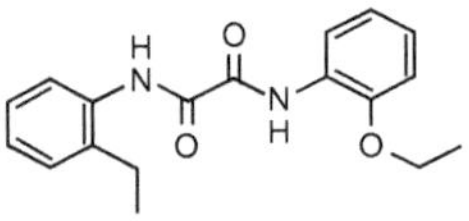	S-[CH_2-CH_2-COO-$(CH_2)_{11}$-CH_3]$_2$
Tinuvin 312 = UV absorber for PEST and PC, N-(2-ethoxyphenyl)-N′-(2-ethylphenyl)-ethylene diamide CAS#: 23949-66-8; $C_{18}H_{20}N_2O_3$ MW: 312.37 g/mol; T_m = 124–127°C, T_b > 400°C	**Irganox PS800** = thiosynergist antioxidant, dilauryl thio-*di*-propionate, DLTDP CAS#: 123-28-4; $(C_{15}H_{29}O_2)_2S$ MW: 514.86 g/mol; T_m = 39.5–41.5°C, T_b = 580.8°C

contain 0.05–0.10 wt% of antioxidants. The AO1 are radical scavengers or H-donors, e.g., Irganox 1010, while the long-term protection is provided by AO2, e.g., hydroperoxide decomposers such as Irgafos 168. Byproducts of the transformation and degradation reactions, as well as additives and pigments, are potentially leachable and extractable.

Several books and other publications on the stabilization of polymers have been mentioned in this chapter. In addition, two patents are noteworthy. The one by Hamilton discusses protection of polymer blends against transesterification.[154] For example, blends of PC with semi-crystalline PEST (e.g., PET) finds application in the automotive industry. Since the resistance to solvents depends on the PET crystallinity, which decreases with advancing transreactions, its control is essential. The patent specifies addition of a silyl phosphate compound for inhibiting the ester-carbonate interchange. The stabilization is achieved by deactivation of the residual metallic catalyst in the reaction with silyl phosphate stabilizers or their mixture.

The second patent addresses several types of degradation mechanism, of the engineering or specialty polymers.[155] Thus, oxidative, thermal and/or

Table 2.6 Examples of labile hydrogens[152]

Polymer	Labile hydrogen, **H**
Polyamide	–CO–NH–C**H**$_2$–
Polyester	–CO–O–C**H**$_2$–
Polyacetal	–O–C**H**$_2$–O–
Unsaturated system	–CH=CH–C**H**$_2$–
Polypropylene/polyethylene	–CH$_2$–C**H**R–CH$_2$–

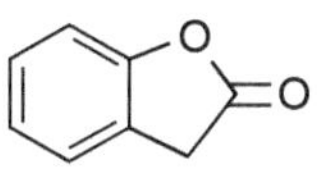

2.13 Benzofuran-2-one.

light-induced degradation of PC, PEST, polyketones or their blends may be prevented by addition of substituted benzofuran-2-one derivatives (see Fig. 2.13). In principle, the function of these molecules is radical scavenging; thus the patent also provides long lists of antioxidants, UV absorbers, light stabilizers, metal deactivators, phosphites and phosphonites, hydroxylamines, nitrones, thio-synergists, peroxide scavengers, co-stabilizers, etc. Nine examples offer stabilizing formulations for PC, PET and polybutylene terephthalate (PBT).

2.8 Challenges in blending polymers

The principal challenges in blending polymers are:

- potential variability of morphology;
- possible reduction of thermal stability;
- incompatible additives in blended polymers.

These problems are discussed below.

Since PAB performance depends on morphology, great efforts are expended to ascertain that the produced articles have optimized structure. However, since the molecular mobility exists above the *beta*-transition temperature, $T > T_\beta \approx 0.86\ T_g$,[156] the morphology may change during the lifetime of the article, especially if exposed to high T. Fortunately, there are several methods of morphology stabilization, including compatibilization, modification of the interphase, crosslinking, incorporation of solids (fillers or crystallization), the use of self-reinforcing miscible systems, etc.

The degradability and stabilization of HTPBs was discussed in Sections 2.6 and 2.7. Now, the influence of one macromolecular blend component on another is of interest. Evidently, in totally immiscible blends the inter-species influences are expected to be small and the blend degradability should be an additive function of composition. By contrast, the largest effects should be observed for miscible blends, for example, transreaction in condensation polymer blends, and free radical dominated processes in the addition polymers with –C–C– bonds in the main chain. At low conversion the transesterification improves compatibility of the condensation polymers, but at higher ones it reduces crystallinity, mechanical performance and thermal stability, as well as the solvent and chemical resistance.

Evidently, degradability depends on many variables (including composition), but from the limited perspective of mutual polymer-polymer effects it may be stated that transreactions, which reduce MW and crystallinity, are detrimental. However, the free radical reactions may have stabilizing effects through grafting, compatibilization and crosslinking. The presence of unsaturated bonds may be beneficial by their free radical scavenging/entrapment capability. Frequently the effect of polymer-1 on polymer-2 is opposite to that of polymer-2 on polymer-1, as reported, for example, for blends of PVC with PMMA or with polybutadiene rubber (BR). However, as the degradation mechanism depends on many variables (presence of O_2, T, P, additives, concentrations, contaminants, etc.) the additivity should not be assumed and each blend should be tested – see the cited reviews for examples.[157–161]

A few examples of the effect of polymer-1 on the degradation of polymer-2 are presented in Table 2.7. Notably, during the dynamic TGA scans the degradation process passes through several stages, each characterized by its own set of chemical reactions. The accelerated and reduced degradation effects have been reported. In blends the less stable component degrades first, either by branching and crosslinking, or chain scission with fragmented byproducts, which may affect stability of the second polymer. Thus, in PC/PVAc or PC/PVC deacetylation and dehydrochlorination liberate acetic or hydrochloric acid, respectively, which accelerates PC hydrolysis.

Most of the blends listed in Table 2.7 were prepared using commercial polymers or blends; thus the results also contain the effects of industrial additives. Furthermore, blend compounding for specific morphology and performance affects blend degradability by consumption of stabilizer, chain scission, addition or formation of compatibilizers, etc. Thus, variability of the degradative behavior for the same polymer-1/polymer-2 mixture is to be expected.

There are thousands of chemical species used as additives, namely antioxidants, metal deactivators, flame retardants, colorants, plasticizers, impact modifiers, lubricants, fungicides, bio-stabilizers, nucleating agents, antistatic agents, crosslinkers, etc. In addition, there are contaminants left after polymerization, compounding or foaming processes. Ideally, the additives should be miscible with the polymer they are supposed to reinforce, protect or color. However, thermodynamics requires that the low molecular

Table 2.7 Examples of polymer-1 degradative effects on polymer-2 in HTPBs

Year	Blend	Effect	Comment	Ref.
1988	PC/PBT	Embrittlement caused by MW reduction and PBT crystallization; transreaction not observed	Aged 42 days at 130°C	162
1991	PC/PBI	PBI catalyzes degradation of PC	17 h at 250°C	163
1997	PMMA/PSS	Polystyrene sulfides (PSS) stabilize PMMA	TGA up to 450°C	164
1998	PPEK/LCP	LCP lowers the degradation *T* of poly(phenolphthalein-ether-ether ketone) (PPEK or PEK-C)	Immiscible blends; TGA up to 600°C	165
2000	LDPE/PPZ	Polyphosphazene increased the LDPE crystallinity (by 11%) and the maximum degradation *T* (by 14°C)	Immiscible blends; TGA up to 800°C	166
2000	PA-6/PP	PP and PP grafted with *di*-ethylmaleate (DEM) reduced the degradation *T* by 60 and 40°C, resp.	Immiscible blends; TGA up to 600°C	167
2001	PA-6/PC/PPG	PA-6/PC transreaction slightly improved stability of the PA-6/PC and PA-6/PC/PPG blends	Immiscible blends; TGA up to 700°C	168
2003	*s*-PS/PPE	In blends the syndiotactic PS (*s*-PS) degrades at higher *T* than neat, while PPE just the opposite	Miscible, crystalline blends; TGA to 700°C	124
2003	PEEK/PAES	Blends degrade faster than expected by assuming additivity	TGA up to 800°C	169
2004	PC/PVC	In blends PC degrades significantly faster than neat, while PVC shows slight stabilization	Miscible blends; TGA up to 600°C	170
2005	PA-6/PP, PA-6/EPDM, PP/TPU	Addition of a compatibilizer, PP-grafted with maleic anhydride (PP-MAH), little affects the degradation	Immiscible blends; TGA up to 600°C	171
2006	PEEK/LCP	Addition of LCP accelerates degradation of PEEK. By contrast, a small stabilization effect is noted for LCP.	Immiscible blends; TGA up to 700°C	172

(*Continued*)

Table 2.7 Continued

Year	Blend	Effect	Comment	Ref.
2006	PC/PVAc	At $T \geq 300$°C deacetylation of PVAc accelerates degradation of PC well above that of neat polymer	Immiscible blends; TGA, pyrolysis + MS to 650°C	173
2007	MDMO-PPV/ $PC_{60}BM$	Presence of $PC_{60}BM$ reduced degradation of MDMO-PPV in the presence or absence of O_2	Photodegradation at 60°C for up to 2000 min	174
2007	PSPI/PBA-a	Polysiloxane-*b*-polyimide increases T_g and char of polybenzoxazine and lowers the activation energy, E_a	Immiscible blends; TGA up to 1000°C	175
2008	PA-12/PP	For w(PA) ≥ 50 and w(PA) ≤ 30 wt%, degradation was similar to that of the matrix; addition of PP-MAH reduced blend degradation	Immiscible blends; TGA up to 700°C	176
2010	PEEK/PEI	Thermo-oxidation of PEEK is enhanced by PEI. The highest E_a is for 50:50 composition – recommended	Miscible blends; TGA up to 1100°C in air or Ar	177
2010	PHB/S-PET	Blending reduced degradability of poly(3-hydroxy butyrate) and that of sulfonated-PET	Miscible and immiscible blends; TGA up to 900°C	178

weight species migrate to the interphase in immiscible PAB. Since miscibility depends on concentration, concentrated additives may precipitate within the interphase, reducing their functionality.

When blending semi-crystalline type PA and/or PEST polymers, information about polymeric chain terminal groups is essential. For example, unprotected chain ends of one polymer may cause transreaction, which may transform the blend into an amorphous copolymer, usually with reduced MW. Different types of stabilizers are needed for different polymers, but when combined they may react and neutralize each other. Similarly, the presence of hydrophilic inorganic stabilizers in one polymer may hydrolyze ester groups of others stabilizing molecules, e.g., phosphites, phosphonites or hydroxy-phenol esters.

2.9 Summary

This chapter describes high temperature polymer blends (HTPBs), methods for their characterization and how these can be used to understand blend functionality. By adjusting the molecular parameters and incorporating suitable additives, commercial polymers may be formulated for a variety of applications. Dozens of different additive types (e.g., heat, light stabilizers, flame retardants, pigments, lubricants, fungicides, nucleating agents, catalyst deactivators, metal scavengers, etc.) amount to 9–15 wt% of the total content. Furthermore, the polymer (e.g., PA or PEST) may be impact modified with up to 20 wt% elastomer. However, in spite of the biphasic nature of the compound it is not labeled as a blend. It seems that with time the number and diversity of blends increase while the number of systems described as blend decrease.

Characterization methods for analyzing blends may be divided into two types: single-phase and multi-phase. Evidently, the number of single-phase systems is limited to amorphous polymers with miscible additives. The semi-crystalline polymers (such as PA-6 or PET) are suspension of the crystalline phase in a vitreous or molten phase, thus subjected to the same analysis of phase formation and evolution with stress and temperature as that of an immiscible blend. The characterization of single-phase systems focuses on the individual macromolecules, their configuration, conformation, molecular weight and its distribution, as well as on properties directly related to the molecular mass and constitution, namely, stability, thermodynamic interactions, rheology, etc.

The performance of multiphase systems is affected by the same parameters as that of a single-phase system, but in addition by the interphase, the global morphology and that of individual phases, as well as the structure variation with processing and forming parameters. Since industrial polymers contain different additives, one must also consider their effects. For example, we blended under the same conditions a series of PP with PA-6 polymers from different manufacturers and containing different additives. In one case a chemical reaction between incompatible additive cocktails resulted in rapid degradation of the blend, while using the same resins with different polymers resulted in respectably performing blends.

Another set of the characterization tests serves for optimization of a single- or multiphase system composition and processing. The characterization may involve mechanical, thermal, physical, flammability, electrical and other tests, which either are directly applicable to any polymeric compound or their applicability extended to the conditions of interest to HTPBs. However, since these materials are often used at high CUT where molecular dynamics affect the blend morphology, the verification of long-term performance may be required. Again, the presence of foreign ingredients, brought in by blending, part mounting or combination of the degrading influences (e.g., temperature and radiation) need to be examined and appropriate stabilizers implemented.

A very special category of characterization methods involves the use of HTPBs in highly demanding engineering applications such as, for example, structured membranes for gas or liquid separation at high temperature, *T*, and pressure, *P*, the fuel cell membranes and active layers in solar cells. Even more demanding are aerospace applications, where the polymers may be exposed to wide *T* and *P* variations, hard radiation and erosion by atomic oxygen. As discussed in Chapter 3, evaluation of the material performance for these applications requires elaborate specific test instrumentation and methods, often developed and maintained by a national organization.

2.10 Future trends

During the last half century polymer blends, like filled polymers before, became indispensable. Combining polymers with desired properties into new materials reduces developmental cost and expense for production and storage of numerous grades. PAB seldom offers synergistic properties, but provides the possibility of creating new material with a combination of properties rarely accessible from a single polymer.[4,122]

On the basis of the historical perspective, the future evolution of PAB technology will be motivated by three principal vectors:

1. new application and requirements;
2. new available materials;
3. new procedures and technologies.

In the case of the first of these vectors, some elements of this future HTPB technology are mentioned in this chapter, e.g., applications in aerospace, fuel and solar cells. For example, the use of polymers and their blends in the membrane application is in a relatively early stage, expensive, lacking efficiency and durability. For fuel cells in particular there are intense efforts for replacing expensive Pt catalyst and the polymeric per-fluoro sulfonic acid (Nafion®). Highly promising developments of these materials have been announced.[179–181] The others include expanded use of PAB in electronic (e.g., as organic transistors) and optoelectronic (photovoltaics, LED) applications, as conductive material (plastic wires), bio-materials and others.[182,183]

Regarding the second vector, to this category belong new polymers and new additives. Some of these have been mentioned, for example, new high performance polymers, self-reinforcing miscible blends, self-organizing macromolecules, etc. Another group of materials that belongs to this category are nanoparticles, which with growing frequency are incorporated into PAB, alone or in combination with more traditional micro-reinforcements. The clay platelets migrate either to the interphase or to a more polar one. Their presence increases the modulus, reduces permeability by gases and vapors, and reduces and stabilizes the size of the dispersed phase.[184,185] The nano-particle technology expanded to fuel and solar cell membranes. As in other fields, replacement of nano-size spheres by rods and plates offers significant improvement at low concentration.[183,186] In addition, nanoparticles influence crystallization of the semicrystalline polymer in its vicinity.[187] Layered clay nanoparticles, layered graphene, carbon nanotubes, metallic or metal oxide nanoparticles are used industrially. Hopefully, collaboration between the ceramic and polymer industries will yield a new family of synthetic, high temperature resistant nanoparticles with affinity to specific polymers that will eliminate the need for intercalation or compatibilization.[188–190]

Regarding the third vector, during the last few years there has been significant progress in the synthetic methods that lead to formation of self-organized macromolecular structures, based on the concept of end-groups attractive/repulsive interactions or formation of block copolymers, namely HyperMacs and HyperBlocks.[191–193] Control of chain ends also offers means for efficient control of the surface properties, e.g., addition of $\leq 0.5\%$ of multi-functional fluoroalkyl additives transforms PS surface tension to that of PTFE.[194]

At the same time computational methods based on the molecular mechanics, molecular dynamics (MD), Monte Carlo (MC) and *ab initio* methods evolved to the high level of predictive capability, accelerating development of new materials with designed properties.[61,195] For the multicomponent and multiphase systems (e.g., nanocomposites, blends, composites with properties adjusted by incorporation of intercalants or compatibilizers) the most promising is the hierarchical multi-scale molecular modeling (MSMM).[196] Thus, while such molecular simulations as MD or MC efficiently analyze the molecular structure at 0.1–10 nm, MSMM, using for example the dynamic mean field density functional theory and dissipative particle dynamics, expands to mesoscopic level larger by a factor of 1000, where the systems morphology may be computed. In the next, macroscopic scale, the finite elements modeling may be used. The model is capable of predicting macroscopic properties of a part, such as degradability, mechanical performance, etc.

2.11 References

1. Parkes Alexander of Birmingham, Artist in the County of Warwick, Brit. Patent, 1,147, 25 Mar 1846; Brit. Patent, 1,313, 07 November 1865.
2. Utracki L. A. and Weiss R. A., Eds., Multiphase Polymers: Blends and Ionomers, ACS Symposium Series No. 395, Washington, DC (1989).
3. Utracki L. A., *Polymer Alloys and Blends*, Hanser Verlag, Munich, New York (1989); revised version, translated into Japanese by Prof. T. Nishi was published by Tokyo Kagaku Dozin Co. Ltd., Tokyo (1991).
4. Utracki L. A., Volume and Series Ed., Two-phase polymer systems, Vol. 2 in *Polymer Processing Society Progress in Polymer Processing Series*, Hanser Publications, Munich (1991).
5. Utracki L. A., *Commercial Polymer Blends*, Chapman & Hall, London (1998).
6. Utracki L. A., Ed., *Encyclopaedic Dictionary of Commercial Polymer Blends*, ChemTec Publishing, Toronto, Canada (1994).
7. Utracki L. A., Ed., *Polymer Blends Handbook*, 2 volumes, Kluwer Academic Publications, Dordrecht (2002).
8. Isayev A. I., Ed., *Encyclopedia of Polymer Blends*, 5 Volumes, Wiley-VCH, Weinheim (2010–2014). ISBN-10: 3-527-31928-X; ISBN-13: 978-3-527-31928-2.
9. http://www.pardos-marketing.com/paper_g04.htm.
10. Niessner N.; Seitz F.; Fischer W.; Guentherberg N.; Ruppmich K.; Moors R. and Weiss R., Eur. Patent Appl., 576,960, 5 January 1994; Ger. Patent, 4,221,293, 5 January 1994; Jap. Patent, 60 57,154, 1 March 1994, Appl. 29 June 1992, to BASF AG.
11. Laughner, M. K., U.S. Patent, 5,068,285, 26 Nov 1991; PCT Int. Appl., WO 93 19,128, 30 September 1993; US Patent, 5,262,476, 16 November 1993; US Patent, 5,270,386, 14 December 1993; US Patent, 5,286,790, 15 February 1994; U.S. Patent, 5,308,894, 03 May 1994, to Dow Chem. Co.
12. http://www.dyneemamatters.com/HB80/.
13. Morin B. G., Polypropylene fiber for reinforcement of matrix materials, US Patent, 7,704,595, 27 April 2010, to Innegrity, LLC (Greer, SC).
14. http://www2.dupont.com/Kevlar/en_US/uses_apps/helmets.html.
15. Marrocco III, M. L.; Gagne, R. R.and Trimmer, M. S., Rigid-rod polymers, US Patent, 5,565,543, 15.10.1996, to Maxdem Inc. (CA).
16. Thomas, D. B.; Maljkovic, N.; Schuler, T.; Rushing, T. S. and Carter R. L., US Patent. Applications, 2009/0069511 A1, 12.03.2009, to Solvay Advanced Polymers, LLC.
17. Utracki L. A., Introduction to polymer blends, Chapter 1 in *Polymer Blends Handbook*. L. A. Utracki, Ed., Polymer Blends Handbook, Kluwer Academic Publications, Dordrecht (2002).
18. Sauer, B. B.; Hsiao, B. S.and Faron, K. L., Miscibility and phase properties of poly(aryl ether ketone)s with three high temperature all-aromatic thermoplastic polyimides, *Polymer*, **37** (3), 445–453 (1996).

19. Fan, Z., Hyperbranched aromatic polyesters and their application in blends of linear polyamides, Doctor rerum naturalium dissertation, Technischen Universität, Dresden (August, 2009).
20. Jaffe, M.; Chen, P.; Choe, E.-W.; Chung, T.-S. and Makhija, S., High performance polymer blends, *Adv. Polym. Sci.*, **117**, 297–327 (1994).
21. Wang, Y.-F.; Hsu, T.; Hay, A. S.; Li, K.and Patel, B., High temperature polymer blends of poly(aryl ether ketone phthalazinone), US Patent Application, 20110104417 A1, 05.05.2011, to Polymics, Ltd., State College, PA (USA).
22. Dai, Y.; Jian, X.; Zhang, S. and Guiver, M. D., Thermostable ultrafiltration and nanofiltration membranes from sulfonated poly(phthalazinone ether sulfone ketone), *J. Membr. Sci.*, **188** 195–203 (2001).
23. Zheng, L.; Liao, G.; Gu, T.; Qu, M.and Jian, X., Mechanical properties of continuous carbon fiber reinforced PPESK blends composites, *Acta Materiae Compositae Sinica*, **25** (3), 1–7 (2008).
24. Yi, Z.; Xu, Y.-Y.; Zhu, L.-P.; Dong, H.-B.and Zhu, B.-K., Hydrophilic modification of PPESK porous membranes via aqueous surface-initiated atom transfer radical polymerization, *Chinese J. Polym. Sci.*, **27** (5), 695–702 (2009).
25. Lee, S., Extreme performance – or processability? new tp polyimide offers both, Plastics Technol., Jan. 2007; (http://www.ptonline.com/articles/200701fa6.html).
26. Sheth, K. C. 2008. Polyetherimide-polyetheretherketone blends and miscibility, ANTEC 2008, *Proceedings of the 66th Annual Technical Conference & Exhibition,* Milwaukee, WI, 4–8.05.2008, Society of Plastics Engineers, pp. 890–894.
27. Sheth, K. C.; R. Gallucci, R. R.and Haralur, G., Thermoplastic polyimide – polyetheretherketone blends with exceptional high temperature properties, ANTEC 2008, *Proceedings of the 66th Annual Technical Conference & Exhibition*, Milwaukee, WI, 4–8.05.2008, Society of Plastics Engineers, pp. 794–797.
28. Sheth, K. C., Highest heat amorphous thermoplastic polyimide blends, ANTEC 2008, *Proceedings of the 66th Annual Technical Conference & Exhibition*, Milwaukee, WI, 4–8.05.2008, Society of Plastics Engineers, pp. 1804–1807.
29. Sabic, October, 2007: Introducing a new level of high performance, high temperature materials, VICTREX MAX-Series polymer; http://www.sabic-ip.com/gep/en/NewsRoom/NewsArticleDetail/ october_2007_victrexmaxseriespolymers.html.
30. Chakravarti, S.; Schrauwen, B. A. G.and van de Grampel, R. D., Polyester polycarbonate compositions, methods of making, and articles formed therefrom, US Patent, 7,829,632, 09.11.2010, to SABIC Innovative Plastics IP B.V.
31. Bhandari, Y. J.; Gallucci, R. R.; Sanner, M.and Singh, R. K., Polyetherimide-polyestercarbonate blends with low heat release‖, ANTEC 2008, *Proceedings of the 66th Annual Technical Conference & Exhibition*, Milwaukee, WI, 4–8.05.2008, Society of Plastics Engineers, pp. 1791–1794.
32. Sanner, M. A.; Gallucci, R. R.and May, A., Injection molding of transparent high impact polyetherimide-polyester carbonate blends‖, ANTEC 2008, *Proceedings of the 66th Annual Technical Conference & Exhibition*, Milwaukee, WI, 4–8.05.2008, Society of Plastics Engineers, pp. 1278–1282.
33. Tian, J.; Kim, S. D.and Chakravarti S., Phase behavior of blends containing poly(resorcinol phthalate-block-carbonate) and poly(ethylene terephthalate)‖, ANTEC 2009, *Proceedings of the 67th Annual Technical Conference & Exhibition*, Chicago, IL, 22–2406.2009, Society of Plastics Engineers, pp. 1438–1442.

34. Naitove, M., Extreme performance copolymers push the boundaries of polycarbonate, Plast. Technol., May 2010; http://www.ptonline.com/articles/'extreme-performance'-copolymers-push-the-boundaries-for-polycarbonate.
35. Bhandari, Y. J.; Sheth, K. C.and Haralur, G., Novel aromatic polyketone blends with a new siloxane polyetherimide copolymer, ANTEC 2010, *Proceedings of the 68th Annual Technical Conference & Exhibition*, Orlando, FL, 16-20.05.2010, Society of Plastics Engineers, pp. 895–898.
36. Haralur, G. M.; Kailasam, G.and Sheth, K., Aromatic polyketone and polysiloxane/polyimide block copolymer composition, US Patent Application, 20090234060 A1, 17.09.2009, to Sabic Innovative Plastics.
37. Utracki, L. A., *Clay-Containing Polymeric Nanocomposites*, book 786 p, RAPRA, Shawbury, Shrewsbury, Shropshire UK (2004).
38. DSM Dyneema Launches Dyneema® HB80 UD material providing highest protection at lightest weight in armor applications, DSM Dyneema Press Release DYNPR141EN0609, 25.06.2009. http://www.dyneemamatters.com/HB80/files/DSM%20DYNEEMA%20LAUNCHES%20DYNEEMAR%20HB80.pdf
39. Morin, B. G.; Adams, B.; Follo, B.and Salem, D., Weight reduction and cost savings using hybrid composites containing high modulus polypropylene fiber, American Composites Manufacturers Association, COMPOSITES & POLYCON 2009 Tampa, FL USA, January 15–17, 2009; Morin, B. G., Polypropylene fiber for reinforcement of matrix materials, US Patent, 7,704,595, 27 April 2010, to Innegrity, LLC (Greer, SC).
40. Bureau International des Poids et Mesures. (2005), 'What is metrology', Copyright BIPM 2004, [Online] http://www.bipm.org/en/bipm/metrology/; BIPM, International Vocabulary of Metrology – Basic and General Concepts and Associated Terms, VIM, 3rd edition, JCGM 200 (2008).
41. Leach R., Fundamental Principles of Engineering Nanometrology (Micro and Nano Technologies), 352 pp., William A., National Physical Laboratory (NPL), UK. (2009) ISBN: 0080964540
42. Czichos, H.; Saito, T. and Smith, L., Eds., *Springer Handbook of Materials Measurement Method*, XXVI, 1208 pp, 970 figures, Springer, Heidelberg, Germany (2006).
43. Utracki, L. A.; Broughton, W.; Gonzalez Rojano, N.; Hecker de Carvalho, L.and Achete, C., Clays for polymeric nanocomposites, *Polym. Eng. Sci.*, **51**(3), 559–572 (2011); Article first published online: 10 January 2011; DOI: 10.1002/pen.21807.
44. Stout, K. J.; Sullivan, P. J.; Dong, W. P.; Maisah, E.; Luo, N.; Mathia, T.and Zahouani, H., The development of methods for the characterization of roughness in three dimensions, Report 15178 (European Commission, Brussels, 1993).
45. http://www.enplanet.co.kr/database/enplatech/testmethod.html; http://www.scribd.com/doc/49179731/ASTM-Polymer-and-Plastic-Test-Techniques.
46. http://www.jiyiyou.com/gb/chemical%20resistance%20plastics.html.
47. http://www.curbellplastics.com/technical-resources/pdf/chemical-resistance-plastics.pdf.
48. Silberberg, A.and Kuhn, W., Miscibility of liquids influenced by rate of shear, *Nature*, **170**, 450 (1952).
49. Utracki, L. A., Investigation of the phenomena of coacervation: Part III, Phase equilibrium in the three component system: toluene – ethanol – polystyrene, *J. Appl. Polym. Sci.*, **6** (22), 399–403 (1962)

50. Koningsveld, R.; Kleintjens, L. A. and Onclin, M. H., Thermodynamics of mixed polymer melts, *J. Macromol. Sci.-Phys.*, **B18**(3), 363–376 (1980).
51. Lee, M. H.; Fleischer, C. A.; Morales, A. R.; Koberstein, J. T. and Koningsveld, R., The effect of end groups on thermodynamics of immiscible polymer blends. *II Cloud Point Curves, Polymer*, **42**(21), 9163–9172 (2001).
52. Shultz, A. R. and Young, A. L., DSC on Freeze-Dried Poly(methyl methacrylate)-Polystyrene Blends, *Macromolecules*, **13**, 663–668 (1980).
53. Stanzione, J. F. III; Strawhecker, K. E. and Wool, R. P., Observing the twinkling fractal nature of the glass transition, *J. Non-Crystalline Solids*, **357**, 311–319 (2011).
54. McBrierty, V. J. and Packer, K. J., *Nuclear Magnetic Resonance in Solid Polymers*, Cambridge University Press, Cambridge (1993).
55. Paul, D. R. and Barlow, J. W., A binary interaction model for miscibility of copolymers in blends, *Polymer*, **25**, 487–498 (1984).
56. Utracki, L. A., Thermodynamics of Polymer Blends Chapter 2 in *Polymer Blends Handbook*, 2 volumes, Utracki L. A., Ed., Kluwer Academic Publications, Dordrecht (2002), pp. 123–201.
57. Utracki, L. A., Statistical thermodynamics evaluation of solubility parameters and polymer miscibility, *J. Polym. Sci. Part B: Polym. Phys.* **42**: 2909–2915 (2004).
58. Tiller, A.R. and Gorella, B., Prediction of polymer miscibility from molecular mechanics calculations, *Polymer*, **35,** 3251–3259 (1994).
59. Case, F.H. and Honeycutt, J.D., Will my polymers mix?: methods for studying polymer miscibility, *Trends Polym. Sci.*, **2**(8), 259–266 (1994).
60. Case, F.H. Applications of modeling in polymer-property prediction, *J. Computer-Aided Mater. Design*, **3**, 369–378 (1996).
61. Kim, K.; Utracki, L. A.and Kamal, M. R., Numerical simulation of polymer nanocomposites using a self-consistent mean-field model, *J. Chem. Phys.*, **121** (21), 10766–10777 (2004).
62. Jawalkar, S. S.; Nataraj, S. K.; Raghu, A. V. and Aminabhavi, T. M., Molecular dynamics simulations on the blends of poly(vinyl pyrrolidone) and poly(bisphenol-A-ether sulfone), *J. Appl. Polym. Sci.*, **108**, 3572–3576 (2008).
63. Virnau, P.; Binder, K.; Heinz, H.; Kreer, T. and Müller, M., Molecular simulation of polymer melts and blends: methods, phase behavior, interfaces, and surfaces, in *Encyclopedia of Polymer Blends, Volume 1: Fundamentals*. Isayev, A. I., Ed., 2010, WILEY-VCH Verlag, Weinheim, Germany, pp. 1–44.
64. Schweizer, K. S. and Curro, J. G., PRISM theory of the structure, thermodynamics, and phase transitions of polymer liquids and alloys, *Adv. Polym. Sci.*, **116**, 321–377 (1994).
65. Heine, D. R.; Grest, G. S. and Curro J. G., Structure of polymer melts and blends: comparison of integral equation theory and computer simulations, *Adv. Polym. Sci.*, **173**, 209–252 (2005). DOI: 10.1007/b99431.
66. Graessley, W. W., *Polymeric Liquids & Networks: Dynamics and Rheology*, Garland Science, London (2008).
67. Gusmán, J. D., Schieber, J. D. and Pollard R., A regularization-free method for the calculation of molecular weight distributions from dynamic moduli data, *Rheol. Acta*, **44**, 342–351 (2005).
68. Llorens, J., Rude, E. and Marcos, R. M., Prediction of polymer molecular weight distribution from rheology: polydimethylsiloxane blends, *Mater. Sci. Forum*, 480–481, 281–286 (2005).

69. Borg, T. and Pääkkönen, E. J., Linear viscoelastic models: Part I and II, *J. Non-Newtonian Fluid Mech.*, **156** (1–2), 121–138 (2009).
70. Borg, T. and Pääkkönen, E. J., Linear viscoelastic models: Part III and Part IV, *J. Non-Newtonian Fluid Mech.*, **159** (1–3), 17–25 (2009); 165 (1–2), 24–31 (2010).
71. Utracki, L. A. and Dumoulin, M. M.: Size exclusion chromatography of polyethylenes; reliability of measurements, in Size Exclusion Chromatography, Provder, T., Ed., Ch. 6, pp. 97–112, ACS Symp. Series, 245, Comstock J. M., Series Ed., Washington, DC (1984).
72. Provder, T., Ed., Chromatography of Polymers; Characterization by SEC and FFF, American Chemical Society, Washington (1993).
73. Potschka, M. and Dubin, P. L., Eds., *Strategies in Size Exclusion Chromatography*, American Chemical Society, Washington (1996).
74. Wu, C. S., Ed., Handbook of size exclusion chromatography and related techniques: revised and expanded, CRC Press; 2 edition (4 November 2003).
75. Striegel, A., Ed., *Multiple Detection Size-exclusion Chromatography*, American Chemical Society, Washington, DC (2004).
76. Kurata, M. and Stockmayer, W. H., Intrinsic viscosities and unperturbed dimensions of long chain molecules, *Fortschr. Hochpolym.-Forsch.*, **3**, 196–312 (1963).
77. Mark, H., *Der feste Korper*, Hirzel, Leipzig (1938).
78. Houwink, R., Relation between the polymerization degree determined by osmotic and viscometric methods, *J. Prakt. Chem.*, **157**, 15 (1940).
79. Sakurada, I.; Sakaguchi Y. and Kokuryo, S., Degree of polymerization as determined by intrinsic viscosity, *Chem. High Polym.*, **17**, 227 (1960).
80. Cantow, M. J. R., Ed., *Polymer Fractionation*, Academic Press, New York (1967).
81. Tung, L.H., *Fractionation of Synthetic Polymers*, Marcel Dekker, New York (1977).
82. Francuskiewicz, F. *Polymer Fractionation*, Springer, Berlin (1994).
83. Shirayama, K.; Okada, T. and Kita, S., Distribution of short-chain branching in low-density polyethylene, *J. Polym. Sci., Part A-2*, **3**(3), 907–916 (1965).
84. Wild, L. and Glöckner, G., Temperature rising elution fractionation, in separation techniques, thermodynamics, liquid crystal polymers, *Adv. Polym. Sci.*, **98**, 1–47 (1991), DOI: 10.1007/3-540-53135-1_4.
85. Monrabal, B. and del Hierro, P., Characterization of polypropylene–polyethylene blends by temperature rising elution and crystallization analysis fractionation, *Anal. Bioanal. Chem.*, **399**(4), 1557–1561 (2011), DOI: 10.1007/s00216-010-4061-5
86. Magagnini, P. I.; Paci, M.; Minkova, L. I.; Miteva, Ts.; Sek, D.; Grobelny, J. and Kaczmarczyk, B., Synthesis and characterization of a PE-g-LCP copolymer, *J. Appl. Polym. Sci.*, **60**, 1665–1676 (1996).
87. Zhu, M.; Sun, Y. and Cai, M., Synthesis and characterization of copolymers of poly(aryl ether ketone amide) and poly(aryl ether ketone ketone), *High Perform. Polym.*, **22**, 763–778 (2010).
88. Vandenburg, H. J.; Clifford, A. A.; Bartle, K. D.; Carroll, J.; Newton, I.; Garden, L. M.; Dean, J. R. and Costley, C. T., Analytical extraction of additives from polymers, *Analyst*, **122**, 101R–115R (1997).
89. Smidsrød, O. and Guillet, J. E., Study of polymer-solute interactions by gas chromatography, *Macromolecules*, **2**(3), 272–277 (1969).
90. Al-Saigh, Z. Y. and Munk, P., Study of polymer-polymer interaction coefficients in polymer blends using inverse gas chromatography, *Macromolecules*, **17** (4), 803–809 (1984), DOI: 10.1021/ma00134a047.

91. Etxeberria, A.; Iriarte, M.; Uriarte, C. and Iruin J. J., Lattice fluid theory and inverse gas chromatography in the analysis of polymer-polymer interactions, *Macromolecules*, **28** (21), 7188–7195 (1995), DOI: 10.1021/ma00125a022.
92. Açikses, A. and Aksar, S., Synthesis, characterization and thermodynamic parameters of poly(3-mesityl-2-oxopropyl methacrylate), *Turkish J. Sci. Technol.*, **4**(2), 127–137 (2009).
93. Wu, P.; Deng, K.; Qi, S. and Nie, H., Thermodynamic interactions and characterization of naphthenic oil by inverse gas chromatography, *Phys. Chem. Liquids: An Intern. J.*, 1029–0451, First published on 1 January 2011; DOI: 10.1080/00319104.2010.485264.
94. Reed, W. F. A method for online determination of polydispersity during polymerization reactions, *Macromolecules*, **33** (19): 7165–7172 (2000). doi:10.1021/ma0006023.
95. Frauendorfer, E. and Hergeth, W.-D., Industrial polymerization monitoring, *Macromol. Symp.*, (Special issue: Polymer Reaction Engineering – 10th International Workshop), **302** (1), 1–5 (2011).
96. Alb, A. M.; Drenski M. F. and Reed, W. F., Perspective automatic continuous online monitoring of polymerization reactions (ACOMP), *Polym. Int.*, **57**(3), 390–396 (2008).
97. Alb, A. M. and Reed, W. F., Fundamental measurements in online polymerization reaction monitoring and control with a focus on ACOMP, *Macromol. Reaction Eng.*, **4** (8), 470–485 (2010).
98. Kreft, T. and Reed, W. F. Predictive control of average composition and molecular weight distributions in semibatch free radical copolymerization reactions, *Macromolecules*, **42** (15), 5558–5565 (2009).
99. Zweifel, H., *Stabilization of Polymeric Materials*, Springer, Berlin (1998).
100. Gächter, R. and Müller, H., *Plastics Additives Handbook*, 4th edition, Hanser Publications, Munich (1993).
101. Garton, A.; Carlsson, D. J. and Wiles, D. M., Polypropylene oxidation: the apparent rate constant for peroxy radical termination and the photoinitiation efficiency, *Macromolecules*, **12** (6), 1071–1073 (1979).
102. Stuetz, D. E.; Diedwardo, A. H.; Zitomer, F. and Barnes, B. P., Polymer Flammability. II, *J. Polym. Sci.: Polym. Chem. Ed.*, **18**(3), 987–1009 (1980).
103. Nassar, N.; Utracki, L. A. and Kamal, M. R., Melt intercalation in montmorillonite/polystyrene nanocomposites, *Int. Polym. Process.*, **20**, 423–431 (2005).
104. Rosato, D. V.; Rosato, D. V. and Rosato, M. G., *Injection Molding Handbook*, 3rd edition, Kluwer Academic Publications, Dordrecht (2000).
105. Beyler, C. L. and Hirschler, M. M., Thermal Decomposition of Polymers, Ch. 7 in *The SFPE Handbook Fire Protection Engineering*, 3rd edition, DiNenno, P. J.; Drysdale, D.; Beyler, C. L.; Walton, W. D., Editors, NFPA, Quincy MA (2002).
106. Ito, K.; Kobayashi, Y. and Nanasawa, A., High sensitivity of positron annihilation to thermal oxidation of polyethylene, *Appl. Phys. Lett.*, **82**, 654–656 (2003); DOI: 10.1063/1.1540730.
107. McNeill, I. C. and Rincon, A., Degradation studies of some polyesters and polycarbonates 8. Bisphenol A polycarbonate, *Polym. Degrad. Stabil.*, **31**, 163–180 (1991).
108. Markezich, R. L., Thermally stable polycarbonate compositions, US Patent, 4,335,039, 15.06.1982, to General Electric Co.
109. Cooper, S. M.; Shafer, S. J. and Tyrell, J. A., Stabilized polycarbonate compositions, US Patent, 4,407,996, 04.10.1983, to General Electric Co.

110. Miller, K. F., Hydrolytically stable polycarbonate compositions, US Patent, 4,393,158, 12.07.1983, to General Electric Co.
111. Camino, G. and Sandrucci, P., *Thermogravimetry, in Characterization Techniques and their Application to Polymer Blends*, Simon, G. P., Ed., Oxford University Press (2003), p. 235
112. Inabi, A. and Kashiwagi, T., A calculation of thermal degradation initiated by random scission, unsteady radical concentration, *Eur. Polym. J.*, **23** (11), 871–881 (1987).
113. Han, Y.; Fang, X.-Z. and Zuo, X.-X., Synthesis and properties of novel melt processable isomeric polythioetherimides, *High Perform. Polym.*, **22**, 989–1003 (2010), DOI: 10.1177/0954008310376604.
114. Sojitra, P. N.; Patel, K. C.and Patel, H. S., high performance polyamides based on s-triazine ring: synthesis and characterization, *High Perform. Polym.*, **22**, 974–988 (2010), DOI: 10.1177/0954008310378054.
115. Wang, G.; Geng, Z.; Zhu, X. and Zhang, S Liu, X., Synthesis and characterization of soluble low-κ poly(aryl ether ketone) copolymers with pendent adamantyl groups, *High Perform. Polym.*, **22**, 779–798, (2010), DOI: 10.1177/0954008309358926..
116. Okabe, T. and Morikawa, A., Synthesis and characterization of novel polyimides from 1,1-bis[4-(4-aminophenoxy)phenyl]-1-(4-hydroxyphenyl)ethane, *High Perform. Polym.*, **22**, 799–819, (2010), DOI: 10.1177/0954008309355844.
117. Thiruvasagam, P. and Venkatesan, D., Synthesis and characterization of processable aromatic polyimides, *High Perform. Polym.*, **22**, 682–693 (2010), DOI: 10.1177/0954008309360932.
118. Alam, S. M. M.; Kawauchi, T. and Takeichi, T., Preparation and characterization of rigid polyimide-clay-polydimethylsiloxane hybrid, *High Perform. Polym.*, **22**, 742–760 (2010), DOI: 10.1177/0954008309358469.
119. Naffakh, M.; Ellis, G.; Gómez, M. A. and Marco, C., Thermal decomposition of technological polymer blends 1. Poly(aryl ether ether ketone) with a thermotropic liquid crystalline polymer, *Polym. Degrad. Stabil.*, **66** (3), 405–413 (1999), DOI: 10.1016/S0141–3910(99)00093–2.
120. Rincon, A. and McNeill, I. C., Thermal degradation of polycarbonate-poly(methyl methacrylate) blends by thermal volatilization analysis, *Polym. Degrad. Stabil.*, **18**(2), 99–110 (1987).
121. Lin, Q. and Yee, A. F., Interactions of a liquid crystalline polymer with polycarbonate and poly(ethylene terephthalate), *J. Mater. Sci.*, **32**(15), 3961–3970 (1997).
122. Robeson, L. M., *Polymer Blends; A Comprehensive Review*, Hanser Publishers, Munich, Germany, (2007); p. 358.
123. Mucha, M. and Kryszewski, M., Studies of decomposition of polymer blends by thermogravimetric method, *J. Polymer Sci.: Symposium*, **42**, 1313–1320 (1973).
124. Stack, S.; O'Donoghue, O. and Birkinshaw, C., The thermal stability and thermal degradation of blends of syndiotactic polystyrene and polyphenylene ether, *Polym. Degrad. Stabil.*, **79**, 29–36 (2003).
125. Bacosca, I.; Hamciuc, E.; Bruma, M. and Ronova, I. A., Study of aromatic polyimides containing cyano groups, *High Perform. Polym.*, **22**(6), 703–714 (2010); DOI: 10.1177/0954008309350784.

126. Morlat S., Mailhot, B.; Gonzalez, D. and Gardette, J.-L., Photo-oxidation of polypropylene/ montmorillonite nanocomposites. 1. Influence of nanoclay and compatibilizing agent, *Chem. Mater.*, **16**, 377–383 (2004).
127. Tanoue, S.; Utracki, L. A.; Garcia – Rejon, A.; Tatibouët, J.; Cole, K. C. and Kamal, M. R., Melt compounding of different grades of polystyrene with organoclay: Part 1. Compounding and characterization, *Polym. Eng. Sci.*, **44**, 1046–1060 (2004).
128. Tanoue, S.; Utracki, L. A.; Garcia – Rejon, A.; Sammut, P.; Ton-That, M.-T.; Pesneau, I.; Kamal, M. R. and Lyngaae-Jørgensen, J., Melt compounding of different grades polystyrene with organoclay: Part 2. Rheological properties, *Polym. Eng. Sci.*, **44**, 1061–1076 (2004).
129. Cowie, J. M. G.; McEwen, I. J. and McIntyre, R., Aging and degradation of polymer blends, Ch. 14 in *Polymer Blends Handbook*, Utracki, L. A. Ed., Kluwer Academic Publications, Dordrecht (2002).
130. La Mantia, F. P. and Valenza, A., Thermo-mechanical degradation of polymer blends, Angew. *Makromolekul. Chem.*, **216** 45–65 (1994).
131. La Mantia, F. P. and Mongiovi, C., Thermomechanical degradation of compatibilized and uncompatibilized nylon 6/polypropylene blends, *Polym. Degrad. Stabil.*, **66**, 337–342 (1999).
132. Singh, A and Silverman, J., Eds., Radiation processing of polymers, vol. 2 in PPS Progress in *Polymer Processing*, Utracki, L. A., Series Ed., Hanser Publishers, Munich (1992).
133. van Gisbergen, J. and Overbergh, N., Radiation effects on polymer blends in radiation processing of polymers, Singh, A. and Silverman, J., Eds., vol. 2 in *PPS Progress in Polymer Processing*, Utracki, L. A., Series Ed., Hanser Publishers, Munich (1992).
134. Singh, A. and Bahari, K., Use of high-energy radiation in polymer blends technology, Ch. 11 in *Polymer Blends Handbook*, Utracki, L. A. Ed., Kluwer Acad. Pub., Dordrecht (2002).
135. Rolando, R. J. and Krueger, D. L., Methods of preparing radiation resistant heat sealable polymer blends, US Patent, 5,258,419, 02.11.1993, to Minnesota Mining & Manufacturing, St. Paul, MN
136. Toy, L. T.; Lau, A. N. K. and Leong, C.-W., Polymeric blends, PCT International Applications WO/1994/014890 A1, to Raychem Corporation, Menlo Park, CA (1994).
137. Kobayashi, T. and Tanaka, K., Permselective hollow fibers of polysulfone and their preparation methods. Jpn Patent (Tokkaihei) 4–300636 (1992), pp. 211–219; Canadian Patent, 2064301, to Toray Industries, Inc., Japan, (1992)
138. Cheng, T. C., Kaduk, B. A., Mehan, A. K., Taft, D. D., Weber, C. J. and Zingheim, S. C., Polymeric blends US Patent, 4,935,467, to Raychem Corporation, Menlo Park, CA (1990)
139. Factor, A.; Carnahan, J. C.; Dorn, S. B. and Van Dort, P. C. The chemistry of γ-irradiated bisphenol-A polycarbonate, *Polym. Degrad. Stabil.* **45** (1), 127–137 (1994).
140. Ching, T.-Y., UV Stabilized polycarbonate resins, US Patent, 4,260,732, 07.04.1981, to General Electric Co.

141. Archey, R. L.; Kumpf, R. J.; Meltzer, A. D. and Pielartzik, H., Gamma-radiation-resistant polycarbonate composition, US Patent, 5,399,658, 21.03.1995, to Miles Inc.
142. Krishnan, S. and Ebert, W., Polycarbonate composition resistant to gamma radiation, US Patent, 6,197,854 B1, 06.03.2001, to Bayer AG.
143. Nouh, S. A. and Abutalib, M. M., Effect of fast neutron irradiation on the structural and optical properties of Makrofol DE 1–1 CC polycarbonate, *Radiation Effects and Defects in Solids: Incorporating Plasma Science and Plasma Technology*, **165** (11), 845–856 (2010); DOI: 10.1080/10420150.2010.481672.
144. Mason, J. P., *Proc. Medical Design & Manufacturing West 95*, Anaheim, CA (1995), p. 103.
145. Malinoski, J., Radiation sterilization of polycarbonate-polyestercarbonate copolymers, SPE ANTEC, 2967–2975 (2009).
146. IAEA-TECDOC-1062, Stability and stabilization of polymers under irradiation, January 1999, Final report of a coordinated research project 1994–1997, International Atomic Energy Agency, Industrial Applications and Chemistry Section, Wagramer Str. 5, P.O. Box 100, A-1400 Vienna, Austria (1999).
147. IAEA-TECDOC-1617, Controlling of Degradation Effects in Radiation Processing of Polymers, May 2009, International Atomic Energy Agency, Industrial Applications and Chemistry Section, Wagramer Str. 5, P.O. Box 100, A-1400 Vienna, Austria (2009).
148. Sun, J.-Z.; Zhong, X.-G., Studies on radiation stability of polymers, in stability and stabilization of polymers under irradiation, IAEA, Industrial Applications and Chemistry Section, Vienna, Austria (1999).
149. Alvino, W. M., Ultraviolet stability of polyimides and poly amid-imides, *J. Appl. Polym. Sci.*, **15**, 2123–2140 (1971).
150. Hill, D. J. T.; Rasoul, F. A.; Forsythe, J. S.; O'Donnell, J. H.; Pomery, P. J.; George, G. A.; Young, P. R. and Connell, J. W., Effect of Simulated low earth orbit radiation on polyimides (UV degradation study), *J. Appl. Polym. Sci.*, **58**, 1847–1856 (1995).
151. Zweifel, H.; Maier, R. D. and Schiller, M., *Plastics Additives Handbook*, 6th edition, Hanser Verlag, Munich, Germany (2009)
152. Moeller, H. W., Ed., *Progress in Polymer Degradation and Stability Research,* Nova Science Publ., New York, NY, USA (2008).
153. Asensio, J. A.; Sánchez, E. M.; Gómez-Romero, P., Proton-conducting membranes based on benzimidazole polymers for high-temperature PEM fuel cells. *A Chemical Quest, Chem. Soc. Rev.*, **39**, 3210–3239 (2010); DOI: 10.1039/b922650h.
154. Hamilton, D. G., Polyester-polycarbonate compositions stabilized against ester-carbonate interchange, US Patent, 5,922,816, 13.07.1999, to General Electric Co.
155. Schmitter, A. and Oertli, A. G., Stabilizing polycarbonates, US Patent, 6,310,220, 30.10.2001, to Ciba Specialty Chem. Corp.
156. Utracki, L. A. and Sammut, P., Molten polystyrene structures above the glass transition, T > Tg, *J. Polym. Sci. Part B: Polym. Phys.*, Article first published online: 18 July 2011; **49**, 000–000 (2011); DOI: 10.1002/polb.22313.
157. Grassie, N.; McNeill, I.C. and Samson, J.N.R., Degradation of polymer mixtures-Part 10: The thermal degradation of blends of polyacrylonitrile and poly(methyl methacrylate), *Polym. Degrad. Stabil.*, **1** (1), 17–35 (1979).
158. Wypych, J., Degradation of polymer blends, *Polym. Networks Blends*, **2** (1), 53–64 (1992).

159. Pospšil, J.; Horák, Z.; Kruliš, Z.; Nešpůrek, S. and Kuroda, S.-I., Degradation and aging of polymer blends. I. Thermomechanical and thermal degradation, *Polym. Degrad. Stabil.*, **65** (3), 405–414 (1999).
160. Lomakin, S.M.; Koverzanova, E.V.; Usachev, S.V.; Shilkina, N.G. and Zaikov, G.E., Thermal degradation and combustion of polymeric blends, *J. Appl. Polym. Sci.*, **86** (13), 3300–3311 (2002).
161. Santos, J. M. R. C. A. and Guthrie, J. T., Polymer blends: the PC–PBT case, *J. Mater. Chem.*, **16**, 237–245 (2006); DOI: 10.1039/B502036K.
162. Bertilsson, H.; Franzen, B. and Kubat, J., Ageing of PC/PBT blends, *Plast. Rubber Proces. Appl.*, **10** (3), 137–153 (1988).
163. Musto, P.; Wu, L.; Karasz, F.E. and MacKnight, W.J., Fourier transform infrared spectroscopy investigations of polybenzimidazole/poly(bisphenol-A carbonate) blends, *Polymer*, **32** (1), 3–11 (1991).
164. Ganesh, K.; Latha, R.; Kishore, K.; George, B. and Ninan, K.N., Stabilization of thermal degradation of poly(methyl methacrylate) by polysulfide polymers, *J. Appl. Polym. Sci.*, **66** (11), 2149–2156 (1997).
165. Mi, Y.; Zhenc, S.; Chan, C.-M. and Guo, Q., Blends of phenolphthalein poly(ether ether ketone) and a thermotropic liquid crystalline copolyester, *J. Appl. Polym. Sci.*, **69** (10), 1923–1931 (1998).
166. Chiu, W.-Y.; Wang, F.-T.; Chen, L.-W.; Don, T.-M. and Lee, C.-Y., Crystallization kinetics and thermal degradation behavior of low-density polyethylene blended with poly(bispropoxyphosphazene), *Polym. Degrad. Stabil.*, **67** (2), 223–231 (2000).
167. Gonzalez, J.; Albano, C.; Sciamanna, R.; Ichazo, M.; Rosales, C.; Martnez, J. and Candal, M., Dynamic thermal decomposition of blends of polyamide-6 with functionalized and non-functionalized PP, *Polym. Degrad. Stabil.*, **68** (1), 9–19 (2000).
168. Costa, D. A. and Oliveira, C. M. F., Blends of polyamide-6, polycarbonate, and poly(propylene oxide). II. Reactive compatibilization-thermal degradation relationships, *J. Appl. Polym. Sci.*, **81** (10), 2556–2562 (2001).
169. Nandan, B.; Kandpal, L.D. and Mathur, G.N., Poly(ether ether ketone)/poly(aryl ether sulphone) blends: thermal degradation behaviour, *Europ. Polym. J.*, **39** (1), pp. 193–198 (2003).
170. Sivalingam, G. and Madras, G., Effect of metal oxides/chlorides on the thermal degradation of poly(vinyl chloride), poly(bisphenol A carbonate), and their blends, *Ind. Eng. Chemistry Res.*, **43** (24), 7716–7722 (2004).
171. Roeder, J.; Oliveira, R.V.B.; Becker, D.; Gonalves, M.W.; Soldi, V. and Pires, A.T.N., Compatibility effect on the thermal degradation behaviour of polypropylene blends with polyamide 6, ethylene propylene diene copolymer and polyurethane, *Polym. Degrad. Stabil.*, **90** (3), pp. 481–487 (2005).
172. Naffakh, M.; Gómez, M.A.; Marco, C. and Ellis, G., Kinetic analysis of thermo-oxidative degradation of PEEK/thermotropic liquid crystalline polymer blends, *Polym. Eng. Sci.*, **46** (2), 129–136 (2006).
173. Uyar, T.; Tonelli, A.E. and Hacaloğlu, J., Thermal degradation of polycarbonate, poly(vinyl acetate) and their blends, *Polym. Degrad. Stabil. B* **12**, 2960–2967 (2006).
174. Chambon, S.; Rivaton, A.; Gardette, J.-L. and Firon, M., Photo- and thermal degradation of MDMO-PPV:PCBM blends, *Solar Energy Mater. Solar Cells*, **91** (5), 394–398 (2007).

175. Tiptipakorn, S.; Damrongsakkul, S.; Ando, S.; Hemvichian, K. and Rimdusit, S., Thermal degradation behaviors of polybenzoxazine and silicon-containing polyimide blends, *Polym. Degrad. Stabil.*, **92** (7), 1265–1278 (2007).
176. Jose, S.; Thomas, S.; Biju, P.K.; Koshy, P. and Karger-Kocsis, J., Thermal degradation and crystallisation studies of reactively compatibilised polymer blends, *Polym. Degrad. Stabil.*, **93** (6), 1176–1187 (2008).
177. Ramani, R. and Alam, S., Composition optimization of PEEK/PEI blend using model-free kinetics analysis, *Thermochim. Acta*, **511** (1–2), 179–188 (2010); DOI: 10.1016/j.tca.2010.08.012.
178. Silva, R.; Carvalho, G.M.; Muniz, E.C. and Rubira, A.F., Miscibility influence in the thermal stability and kinetic parameters of poly (3-hydroxybutyrate)/poly (ethylene terephthalate) sulphonated blends, *Polimeros*, **20** (2), 153–158 (2010).
179. Helm, M. L.; Stewart, M. P.; Bullock, R. M.; Rakowski-DuBois M. and DuBois, D. L., A synthetic nickel electrocatalyst with a turnover frequency above 100,000 s–1 for H2 production, *Science*, **333**, 863–866 (2011).
180. Duncan, A. J.; Leo, D. J. and Long, T. E., Beyond Nafion: charged macromolecules tailored for performance as ionic polymer transducers, *Macromolecules*, **41** (21) 7765–7775 (2008); DOI: 10.1021/ma800956v.
181. Gubler, L. and Scherer, G. G., Trends for fuel cell membrane development, *Desalination*, **250**, 1034–1037 (2010); DOI: 10.1016/j.desal.2009.09.101.
182. Matras-Postolek, K. and Bogdal, D., Polymer nanocomposites for electro-optics: perspectives on processing technologies, material characterization, and future application, *Adv. Polym. Sci.*, **230**, 221–282 (2010); DOI: 10.1007/12_2010_49.
183. Hindson, J. C.; Saghi, Z.; Hernandez-Garrido, J.-C.; Midgley, P. A. and Greenham, N. C., Morphological study of nanoparticle-polymer solar cells using high-angle annular dark-field electron tomography, *Nano Lett.*, **11**, 904–909 (2011), DOI. org/10.1021/nl104436j.
184. Utracki, L. A. Rheology of polymer blends, in *Encyclopedia of Polymer Blends*, Chapter 2 in Volume 2, Isayev, A., Ed., Wiley-VCH Verlag, Weinheim, Germany (2011).
185. Fedullo, N.; Sorlier, E.; Sclavons, M.; Bailly, C.; Lefebvre, J.-M. and Devaux, J., Polymer-based nanocomposites: overview, applications and perspectives, *Prog. Org. Coat.*, **58**, 87–95 (2007); doi:10.1016/j.porgcoat.2006.09.028.
186. Ridolfi, G.; Casalbore-Miceli, G.; Geri, A.; Camaioni, N., Possamai, G. and Maggini, M., Organic photovoltaics: towards a revolution in the solar industry, *Proceedings of a Consultants Meeting,* Bologna, Italy, 22–25 March 2004, *Emerging Applications of Radiation in Nanotechnology,* IAEA-TECDOC-1438, pp.71–83 (2004).
187. Yuan, Q. and Misra, R. D. K., Polymer nanocomposites: current understanding and issues, *Mater. Sci. Technol.*, **22**, 742–755 (2006); DOI: 10.1179/174328406X101292.
188. Utracki, L. A., *Clay-Containing Polymeric Nanocomposites*, book 786 p, RAPRA, Shawbury, Shrewsbury, Shropshire UK (2004).
189. Utracki, L. A.; Sepehr, M. and Boccaleri, E., Synthetic, layered nano-particles for polymeric nanocomposites (PNC's), *Polym. Adv. Technol.*, **18**(1), 1–37 (2007); DOI: 10.1002/pat.852.

190. Kim, H.; Abdala, A. A. and Macosko, C. W., Graphene/Polymer Nanocomposites, *Macromolecules*, **43** (16), 6515–6530 (2010); DOI: 10.1021/ma100572e.
191. Hutchings, L. R., DendriMacs and HyperMacs. Emerging as more than just model branched polymers. *Soft Matter.*, **4**, 2150–2159 (2008).
192. Hutchings, L.R.; Dodds, J. M.; Rees, D.; Kimani, S. M.; Wu, J. J. and Smith, E., HyperMacs to hyperblocks: a novel class of branched thermoplastic elastomer. *Macromolecules*, **42**, 8675–8687 (2009).
193. Matsumi, N.; Kagata, A. and Aoi, K., Synthesis of supramolecular solid polymer electrolytes via self-assembly of diborylated ionic liquid, *J. Power Sources*, **195**, 6182–6186 (2010).
194. Hutchings, L.R.; Sarih, N. M. and Thompson, R. L., Multi-end functionalised polymer additives synthesised by living anionic polymerisation – the impact of additive molecular structure upon surface properties. *Polym. Chem.*, **2**, 851–861 (2011).
195. Harton, S. E.; Koga, T.; Stevie, F. A.; Araki, T. and Ade, H., Investigation of blend miscibility of a ternary PS/PCHMA/PMMA system using SIMS and mean-field theory, *Macromolecules*, **38**, 10511–10515 (2005); DOI: 10.1021/ma051595r.
196. Fermeglia, M. and Pricl, S., Multiscale molecular modeling in nanostructured material design and process system engineering, *Computers Chem. Eng.*, **33** (10), 1701–1710 (2009); DOI: 10.1016/j.compchemeng.2009.04.006.

2.12 Appendix: Nomenclature

2.12.1 General abbreviations

AA	acrylic acid (monomer)
ACOMP	automatic continuous online monitoring of polymerization
AFM	atomic force microscopy
AO1	primary antioxidant (e.g., BHT = 2,6-di-tert-butyl-4-methylphenol)
AO2	secondary antioxidant (e.g., phosphor, or sulfur compounds)
ASE	accelerated solvent extraction
ATREF	analytical temperature rising elution fractionation
BHT	2,6-dimethyl-4-tert-butylphenol
CAS	Chemical Abstracts Service, a division of the American Chemical Society
CF	carbon fiber
COMPASS	condensed-phase optimized molecular potentials for the atomistic simulation studies
CPC	cloud-point curve
CRM	certified reference material
CUT	continuous use temperature
DEM	*di*-ethylmaleate

DSC	differential scanning calorimetry
DTA	differential thermal analysis
DTG	differential thermogravimetry
ESC	environmental stress cracking
ESCA	(also known as XPS) X-ray photoelectron spectroscopy
ESCR	environmental stress crack resistance
ESR	electron spin resonance
FPI	fluoro polyimide
FT-IR	Fourier transform infrared spectroscopy
GPC	gel permeation chromatograph (now: size exclusion chromatography, SEC)
HALS	hindered amine light stabilizers (now HAS)
HDT	heat deflection temperature
HR-NMR	high resolution nuclear magnetic resonance
HTP	high temperature polymer
HTPB	high temperature polymer blend
HTS	high temperature sulfone
IGC	inverse gas chromatograph
LALLS	low-angle laser light scattering
LCP	liquid crystal polymer
LEO	low Earth orbit
LSc	light scattering
MA	microbeam analysis
MAE	microwave-assisted extraction
MAH	maleic anhydride (monomer)
MALDI	matrix-assisted laser desorption/ionization
MS	mass spectrometry
MSMM	multi-scale molecular modeling
MW	molecular weight
MWD	molecular weight distribution
NIR	near-infrared spectroscopy
NMR	nuclear magnetic resonance
PALS	positron annihilation lifetime spectroscopy
$PC_{60}BM$	6,6-phenyl C61-butyric acid methyl ester (also: PCBM)
PCM	phase contrast microscopy
PEST	thermoplastic polyesters, e.g., PBT, PET, also TPES
PI	polyimide
PICS	pulse-induced critical scattering
PM	polarized microscopy
PMDA	pyromellitic *di*-anhydride
PNC	polymeric nanocomposite
PO	polyolefin
ppm	parts per million

PRISM	Polymer Reference Interaction Site Model
PTREF	preparative temperature rising elution fractionation
PVT	pressure-volume-temperature
RISM	Reference Interaction Site Model
SANS	small-angle neutron scattering
SALS	small-angle light scattering
SAXS	small-angle X-ray scattering
SEC	size exclusion chromatography
SEM	scanning electron microscopy
SFE	supercritical fluid extraction
SIMS	secondary ion mass spectrometry
SRP	self-reinforced polyphenylene ether
TEM	transmission electron microscopy
TGA	thermogravimetric analysis
TMA	thermo-mechanical analysis
TOA	thermo-optical analysis
TPI	thermoplastic polyimide
TREF	temperature rising elution fractionation
TSC	thermally stimulated current
TSE	twin screw extruder
TVA	thermal volatilization analysis
UV	ultraviolet
WWII	World War II
XPS	X-ray photoelectron spectroscopy
XRD	X-ray diffraction

2.12.2 Polymer abbreviations

AAS, ASA	copolymer of acrylonitrile, acrylate (ester) and styrene
ABS	thermoplastic terpolymer, an acrylonitrile-butadiene-styrene copolymer
CF	cresol-formaldehyde resins (also reinforcing carbon fiber)
COC	cyclic olefin copolymer
EPDM	elastomeric terpolymer from ethylene, propylene, and a non-conjugated diene
ETFE	copolymer from ethylene and tetrafluoroethylene
FEP	fluorinated EPR; tetrafluoroethylene/hexa-fluoro propylene rubber
GP	gutta-percha
HBA	hydroxybenzoate
HDPE	high density polyethylene (ca. 960 kg/m^3)
HIPS	high impact polystyrene

HNA	hydroxynaphthanoate
LDPE	low density polyethylene (ca. 918 kg/m^3)
LLDPE	linear low density polyethylene
MBS	copolymer from methylmethacrylate, butadiene and styrene
NBR	elastomeric copolymer from butadiene and acrylonitrile; nitrile rubber
NR	natural rubber
PA	polyamide; the abbreviation PA is normally followed by a number, a combination of numbers, a letter or a combination of letters and numbers. A single number refers to the polyamide from an α,ω-amino acid or its lactam. A combination of two numbers is often separated by a comma. The first number following the symbol PA indicates the number of methylene groups of aliphatic di-amines, the second number the number of carbon atoms of aliphatic di-carboxylic acids. An *I* stands for isophthalic acid, a *T* for terephthalic acid. For example, co-polyamide from caprolactam, hexamethylenediamine condensed with isophthalic and terephthalic acids is abbreviated as: PA-6IT6, or that from caprolactam, *m*-xylylenediamine and adipic acid as: PA-mXD6, etc.
PA-12	polyamide-12
PA-46	poly(tetramethylene adipamide), also PTA
PA-6	poly-ε-caprolactam
PA-66	poly(hexamethylene diamine-adipic acid), polyhexamethylene-adipamide
PA-6I	poly(hexamethylene isophthalamide)
PA-6IT6	poly(caprolactam-co-hexamethylene diamine-isophthalic and terephthalic acids)
PAB	polymer alloys and blends
PAEK	polyaryletherketone
PAEKP	poly(aryletherketone phthalazinone)
PAES	polyarylethersulfone
PAI	polyamide-imide
PAN	polyacrylonitrile
PAr, PAR	polyarylate, $[\text{-}\phi\text{-}C(CH_3)_2\text{-}\phi\text{-}CO_2\text{-}\phi\text{-}CO_2\text{-}]_n$, amorphous polyester of bisphenol-A with isophthalic and terephthalic acids
PARA	polyaryl amide (aromatic, usually amorphous polyamide)
PB	poly-1-butene, polybutylene; elastic polydiene fiber
PBI	polybenzimidazole, e.g., poly[2,2′-(*m*-phenylene)-5,5′*bis*-benzimidazol]

PBT, PBTP	polybutyleneterephthalate
PC	polycarbonate of bisphenol-A
$PC_{60}BM$	6,6-phenyl C_{61}-butyric acid methyl ester (also: PCBM)
PCT	poly (cyclohexanedimethylene terephthalate)
PCTG	copolymer of cyclohexanedimethanol (≥ 66 mol %), ethylene glycol (≤ 34 mol %) and terephthalic acid
PCTFE	polychlorotrifluoroethylene; also CEM, CFM, CTFE
PDMS	polydimethylsiloxane
PE	polyethylene
PEA	polyetheramide
PEB	polyethylene-p-oxybenzoate
PEBA	thermoplastic elastomer, polyether-block-amide
PEEI	polyesteretherimide
PEEK	polyetheretherketone
PEG	polyethylene glycol
PEI	polyetherimide
PEK	polyetherketone
PEKK	poly(ether-ketone-ketone)
PES	polyethersulfone $[-\phi-SO_2-\phi-O-]_n$
PEST	thermoplastic polyesters, e.g., PBT, PET, also TPES
PET	poly(ethylene terephthalate)
PFEP	copolymer from tetrafluoroethylene and hexa-fluoro propylene; also FEP
PHB	poly-*p*-hydroxybenzoate
PI	polyimide, but also *trans*-1,4-polyisoprene, gutta-percha (UK)
PIB	polyisobutylene
PMB	poly-*p*-methylenebenzoate
PMMA	polymethylmethacrylate
PMS	poly-α-methyl styrene
PnBMA	poly-*n*-butyl methacrylate
PO	polyolefin
POM	polyoxymethylene, polyformaldehyde, polyacetal or 'acetal resin'
POP	polyoxypropylene, usually PPG
PP	polypropylene
PPA	polyphthalamide
PPD-T, PPTA	poly(*p*-phenylene terephthalamide), *Kevlar*™
PPE	poly(2,6-dimethyl 1,4-phenylene ether)
PPEK	poly(phenolphthalein-ether-ether ketone) (PPEK or PEK-C)
PPESK	poly(phthalazinone ether sulfone ketone)
PPG	polypropylene glycol

PPhA	polyphthalamide
PP-MAH	maleic anhydride-modified polypropylene
PPS	polyphenylsulfide
PPSS, PPS-S	polyphenylene sulfidesulfone, polythioethersulfone
PPSU, PSF, PSO	polyphenylene sulfone; polysulfone
PPZ	polyphosphazene, Polyorganophosphazene
PS	polystyrene
PSF	polysulfone, also PSUL, PSU, PSO
PSPI	polysiloxane-*b*-polyimide
PSS	poly(styrene-sulfonate)
PSU	polysulfone $[-\phi-SO_2-\phi-O-\phi-C(CH_3)_2-\phi-O-]_n$
PTEI	polythioetherimide
PTFE	polytetrafluoroethylene (also TFE)
PTMA	polytetramethyleneadipate
PVAc	polyvinyl acetate
PVC	polyvinyl chloride
PVDC	polyvinylidene chloride; also PVC2
PVDF	polyvinylidene fluoride; also PVF2
PVME	polyvinylmethylether
PVP	poly-N-vinylpyrrolidone
SAN	thermoplastic copolymer from styrene and acrylonitrile
SBR	styrene-butadiene elastomer
SEBS	styrene-ethylene/butylene-styrene triblock polymer
S-PET	sulfonated-PET
sPS	syndiotactic polystyrene
TPI	thermoplastic polyimide
UHMWPE	ultrahigh molecular weight polyethylene (over 3 Mg/mol)

2.12.3 Notation

B_{ij}	Huggins-Flory type binary interaction parameters between segments (i and j)
d_d	limit of phase heterogeneity in PAB
E' and E''	storage and loss tensile or flexural moduli
G'	storage shear modulus
G''	loss shear modulus
M_n	number average molecular weight
M_v	viscosity-average molecular weight
M_w	weight average molecular weight
P	pressure
Q	irradiation intensity

$\langle s^2 \rangle^{1/2}$	radius of gyration of coiled macromolecule T temperature
T1 NMR	spin-lattice relaxation times
T_b	boiling point
T_β	beta transition temperature
T_c	crossover temperature
$T_{c,m}$	middle of crystallization temperature
T_{decomp}	initial decomposition temperature
T_g	glass transition temperature
T_m	melting temperature
V	volume or voltage

2.12.4 Greek

ϕ	volume fraction
ΔG_m	free energy of mixing
ΔH_m	heat of mixing
Δl	interphase thickness in PAB
ΔT_g	width of T_g peak at half-height
δ	Hildebrand solubility parameter
η	viscosity
$[\eta]$	intrinsic viscosity
σ	stress

3
Characterization of high temperature polymer blends for specific applications: fuel cells and aerospace applications

L. A. UTRACKI, formerly of the National Research Council, Canada

DOI: 10.1533/9780857099013.70

Abstract: This chapter examines the use of high temperature polymers as membranes, fuel cells and solar cells. Also, it considers the application of such materials for aerospace developments. The methods for characterization and testing of specific high temperature polymer blends for each of these applications are reviewed.

Key words: fuel cells, solar cells, aerospace, characterization, miscible blends, high temperature polymer blends.

3.1 Introduction

High temperature polymer blends (HTPBs) are typically used at $T \geq 140°C$ and maintain at least 25% of their room temperature (RT) properties at 154°C. Considering their expanding applications in such sectors as the military, aerospace, transportation, electronic, healthcare, and oil and gas industries, they need to have good processability, high mechanical performance, chemical resistance, fire retardancy etc.[1] Table 3.1 provides a list of some of the most important HTPBs in order of increasing continuous use temperature (CUT) from 75°C to 540°C. Building on the previous chapter, this chapter looks at the use of HTPBs for membrane, fuel and solar cell and aerospace applications. It reviews ways of characterizing and testing particular polymer blends for these applications.

Table 3.1 High temperature polymer blends (HTPBs) arranged in order of increasing 'continuous use temperature' (CUT)

Polymer	Examples	Temperature (°C)			Blends with	Comment
		CUT	T_m	T_g		
Polyoxymethylene	POM	85–115	160–175	–80	POM, PEST, PPE, PC	Delrin®
Polyphenyleneether	PPE (PPE/PS)	104	257	205–215	PEST, PC, PA, PARA, PCT, POM	With PPE or styrenics
Polyesters	PET, (PBT, PC)	110, (118)	257, (219)	70, (31)	PBT, PEN, Par, POM, PCT, PEA, PPE/PS, PC	Valox®
Polyamides	PA-6, (PA-66)	110, (99)	220 – 225, (255 – 264)	52, (65)	PPE, PA, PARA, POM, LCP, PET, PBT, PAr, PEB, PC	Nylon, Zytel® (aliphatic)
Polycarbonate	PC	116		147–149	PET, PBT, PCT, PCTG, PAr, PPE/PS, POM	Bis-phenol-A polycarbonate, PC
Polyphthalamide	PPA	120–185		310	PA, PPS, PP, PARA, PC, LCP	Amodel®
Polyarylates	PAr	130		188	PEST, PARA, LCP, PA, PBI, PC, PEI, PEK, PEST, Phenoxy, PMB, PPE, PPS	With heteroatom in the main chain
Polyarylene sulfide	PPS	170	280 – 288	194	PPE, PC, PTFE, PSF, PAr, PARA, PI, PAI, PEK, LCP, PAES, PPSS, PDMS, FEP, PCTFE, PTFE, PVDF	Diverse aromatic; Rayton®
Aromatic (amorphous) polyamides, PARA*	PA-6I, (PA-6I6T),	[180]	–, (310), [310]	120, (90), [127]	POM, PEST, PA, PAI, PPS, PES, PphA, PAr, PBI	Excluding LCP-type
Polyphenylene-terephthalamide†	PPTA, PMTA	180	310	90	PBI	Flame resistant coatings

(*Continued*)

Table 3.1 Continued

Polymer	Examples	Temperature (°C)			Blends with	Comment
		CUT	T_m	T_g		
Polysulfone, PSU, polyphenylsulfone,	PPSU, (PSF), [PES], PSO	190, (140 – 150), [204]		190, [230]	PC, PPE, PA, PEST, PI, PAI, PEI, PEK, PEEK, PTFE, PEA	Mindel®, Eviva® Udel®, Radel®,
Liquid crystal polymers	LCP	220 – 271	300 – 350	ca. 120	PA, PPTA, PARA, PAEK, PPE, PI, PAI, PEI, PPS, PEST, PVDC, PVDF, PVF, PSF, LCP, fluorinated polymers	Xydar®, diverse others
Polyaryletherketones, PAEK	PEEK	240 – 310	334	143	PES, PPS, PI, PAI, PEI, PA, LCP, PSF, fluorinated polymers	Victrex®
Polyimides	PAI (PEI)	250 – 275, (165 – 170)		279, (220)	PA, PARA, PEST, PC, PPE, PPS, PSF, PEEK, LCP (PC, PA, PSF, PEC, PAr, PEST, LCP, PPS, PEEI)	Ultem®
Thermoplastic polyimide	TPI	230 [465]	– [388]	≤ 311 [250]	PEI, LCP, PTFE, PAI, PPE, PARA, PPS, PSF, PEEK, PI	Extem®, Aurum®
Polybenzimidazole‡	PBI	435	760	427	PEEK, PPTA, PVDF, PES, PEI, PAI, PARA	High-T blends

Notes: *PA-6I, poly(hexamethylene isophthalamide); †PPTA, poly(p-phenylene terephtalate) and PMTA, poly (m-phenylene terephthalate); ‡PBI, poly[2,2'-(m-phenylene)-5,5'bis-benzimidazol].

3.2 High temperature polymer blends (HTPBs) for membrane applications

During the last 30 years or so membrane applications have seen a significant growth in diverse fields of application. For example, they are used for:

- water, ultrapure water and wastewater purification;[2–5]
- reverse osmosis (RO);
- seawater desalination;
- microfiltration, ultrafiltration and nanofiltration;
- membrane bioreactors for wastewater;
- food, biotechnology and chemical applications;[6,7]
- tissue engineering, biopharmaceutical separations, hemodialysis;
- purification of juice, saccharides, beer and wine, condiments, oil, gluten, etc.;
- vapor and gas separation, for example in the petroleum industry (H_2, CO_2, CH_4, H_2S, SO_2, NO_X, He, H_2O, hydrocarbons and others);[8–11]
- use as contactors and reactors;[12]
- environmental and energy applications;
- fuel cell membranes;[13,14]
- solar cell membranes, etc.[15,16]

Evidently, not all these membrane applications require HTPBs. Water purification by ultra and nanofiltration (UF, NF) does make use of HTPBs. According to the American Membrane Technology Association (AMTA), the pressure in these operations ranges from 0.1 to 63 MPa, operating temperature is 120–180°C and the membrane must survive the temperature drop to −40°C. In consequence, hollow or spirally wound fibers, cartridges, and tubular membranes are prepared, for example from cellulose acetate (CA), semi-aromatic *co*-polyamide (CPARA), polysulfone (PSF) or PVDF. They are used in multi-layered arrangements with blends, which provide desirable gradients of morphologies.

The need for HTPs or HTPBs is often for chemical (e.g., preventing of swelling or oxidation) and mechanical performance. For example, in the petroleum industry dual-layer hollow fiber membranes are used for separation of H_2/CO_2 and CO_2/CH_4 gases. These are prepared from solutions of 1:1 PIB/PI blend (e.g.: poly[2,2'-(1,3-phenylene)-5,5'-bibenzimidazole] and poly[3,3'4,4'-benzophenone-*tetra*-carboxylic-*di*-anhydride-*co*-5(6)-amino-1-(4'-amino-phenyl-1,3-*tri*-methyl-indane)]).[17,18] The spun hollow fibers are coated by a solution of silicone rubber. The membranes had the desired microstructure. Membrane selectivity of 11.11 and 41.81 was obtained for H_2/CO_2 and CO_2/CH_4, respectively. Furthermore, the new membranes showed good resistance toward CO_2-induced plasticization. Camacho-Zuniga *et al.*

developed PSF copolymers based on an equimolar mixture of the rigid/compact naphthalene moiety having T_g = 230°C, stable to 400°C (10 wt% loss in air) and positron annihilation lifetime spectroscopy (PALS) determined free volume cavity diameters from 9.4 to 13.9 nm. Good gas permeability and selectivity were reported for H_2/CH_4, O_2/N_2, and CO_2/CH_4.[19] The membranes were characterized for the morphology using a scanning electron and a field emission microscope, SEM and FEM, respectively. The permeation performance of the hollow fibers was analyzed using the variable-pressure constant-volume method at 35°C and pressure of 1.0 and 0.35 MPa for H_2/CO_2 and CO_2/CH_4, respectively.[20]

The greatest challenge for high temperature membrane technology is in fuel[13,14] and solar[15,16] cell applications. The general information may be found in the cited reviews – more details are presented in Sections 3.3 and 3.5, respectively.

3.3 Fuel cell (FC) membrane applications

There are several types of fuel cells (FCs), where a fuel is transformed into electricity:[21]

- A-FC (alkaline fuel cell);
- PEM-FC (polymer exchange membrane fuel cell) ;
- DM-FC (direct methanol fuel cell);
- PA-FC (phosphoric acid fuel cell);
- MC-FC (molten carbonate fuel cell);
- SO-FC (solid oxide fuel cell).

Most FCs are designed for H_2. A hydrogen-powered FC (HP-FC) typically consists of two electrodes and a PEM separator. DM-FCs are comparable to a HP-FC in regards to operating temperature, but are not as efficient. The fuel is supplied as a 1 mol solution, for enhanced efficiency fed to the cell at high temperatures and pressure. Methanol crossover (≤50% diffuses through PEM without converting to H_2) decreases efficiency. Also, the DM-FCs require a relatively large amount of platinum to act as a catalyst, which makes these fuel cells expensive. There are also attempts to produce FCs fed by hydrocarbons (e.g., natural gas direct carbon fuel cell, NGDC-FC) converting it to H_2O and CO_2.[22,23] NGDC-FC combines methane (CH_4) decomposition with the fuel cell to produce H_2 and electricity. The system is suitable for automotive use, with CH_4 pumps made available at fuel filling stations.

For this chapter, PEMs and DMs, with solid polymer electrolyte membranes, are of interest. Because of different types of fuels (mainly hydrogen or methanol, H_2 or MeOH, respectively) even in this category the materials and construction vary broadly. For example, while for portable applications the PEM

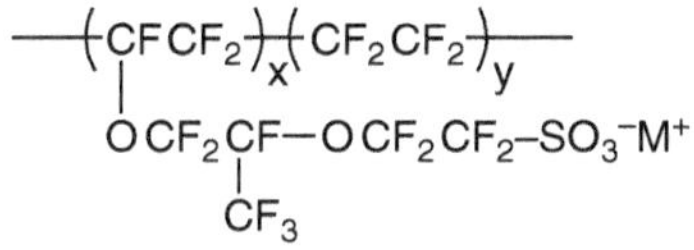

3.1 Nafion®.

will operate at $T < 80°C$, for automotive and stationary applications the T goes to 130°C and 180°C, respectively.[24] For good conductivity of solid PEMs, the presence of water is essential. However, since as T increases keeping water in the PEM gets more difficult, there is a need for reducing the water content to at least 25% without upsetting PEM performance.[25] The H_2 technology remains expensive; in 2006, General Motors estimated that for a FC system to have an equivalent cost to a combustion engine, the FC power plant (PEM +catalyst + fabrication) would need to cost <10$/kW and last 10 years; by 2009 the cost of an 80-kW automotive PEM/H_2 system was $61/kW.[26,27]

The most popular PEM used in fuel cells is *Nafion®*, a sulfonated (thus ionomer) tetrafluoroethylene-based fluoro polymer-copolymer (Fig. 3.1). *Nafion®* is a proton conductor with good chemical, thermal and mechanical stability. However, it is expensive and its performance deteriorates at the operating range for fuel cells, $T > 80°C$. The search for a *Nafion®* replacement is limited by three criteria:

1. the material must be stable at high temperatures;
2. it may be acidified (sulfonated or phosphonated);
3. it may be fluorinated or blended with fluorinated polymers.

Since blends offer a wider spectrum of performances than individual polymers, the search invariably involves HTPBs. In particular, the adopted strategy is using ionomers for electrical conduction and HTP for strength.

Attractive blends for PEMs with high proton conductivity have been made from sulfonated PES, PSU, polyetherketone (PEK), PEEK or poly(2,6-dimethyl 1,4-phenylene ether) (PPE) blended with polybenzimidazole (PBI) or polyetherimide (PEI).[28,29] To preserve the desired PEM performance, the blends are often crosslinked by radiation, chemical reaction of ionic interactions. For long-term PEM applications it is important that membranes' resistance to mechanical, chemical and thermal degradation is maximized. Accelerated aging tests should follow several membrane functionalities, for example conductivity, membrane integrity and permeability.[30] The tests should also identify a possible cross-correlation of effects, namely stress on thermal and/or chemical degradation.

However, PEMs are also being made from polymer blends other than sulfonated polymers; examples are listed in Table 3.2. Since the FC operates

Table 3.2 Some polymer blends used for PEM

No.	Polymer-1	Polymer-2	Comments	Year	Ref.
1	PBI: poly[(m-phenylene)–*bi*-benzimidazole]	PAR: poly[(bis-phenol-A)-(i/t-phthalic acids)]	Miscible via =NH and –COOH interactions	1990	31
2	Sulfonated PSF	PBI or poly ether of pyridine + phenyl phosphine oxide, PPyPO	Doped with phosphoric acid	2003	32
3	Acidic polymer: sulfonated polyether ether ketone (SPEEK)	Basic polymer: poly(vinyl pyridine) (PVP) and a hydrophobic stabilizer	Stabilizer is poly(styrene-b-4vinyl pyridine) with 0–80 wt% styrene	2003	33
4	Amine-terminated sulfone-imide-polyphosphazene	PTFE, PVDF, PVDF-HFP, PS, PBD, PVDC, PMMA, PVAl, PVAc, PPE, PEEK, PET, PBT, PC, PBI, PDMS, PANI	PEM cast from blends of polyphosphazene with other polymer, then crosslinked by γ-rays.	2004	34
5	Acidic polymer: sulfonated PSF, PES, PEEK, PPE,	Basic polymer: PBI, PEI, P4VP, PSF-NH_2, PSF-$N(CH_3)_2$	Blends with or without H_3PO_4 or HCl show high-T performance	2004	35
6	Sulfonated PES	Sulfonated polyphenylene sulfide, SPPS	Solution cast; immiscible particles of SPPS	2004	36
7	Sulfonated poly(ether ketone ketone) (SPEKK)	PES or SPEKK with a different sulfonation level.	Solution-cast PEM had co-continuous morphology	2005	37, 38
8	Sulfonated poly(ether ketone ketone) (SPEKK)	poly(ether imide) (PEI), poly(ether sulfone) (PES) or slightly sulfonated PS particles	Miscibility vs. electrolyte transport depends on miscibility, which is influenced by sulfonation	2006	39, 40
9	SPEKK with ionic exchange capacity: IEC = 1.2	SPEKK with IEC = 2.08; weight ratio = 60/40	A co-continuous blend as a PEM for FC in low-T and at high vapor water	2007	41
10	Sulfonated homopolyimide (SPI)	Sulfonated co-polyimide	SPI conductivity of solvent cast PEM increased with sulfonation	2007	42

11	Sulfonated poly(ether ketone ketone) (SPEKK)	PEI	3:7 blends were aligned in electrical field, increasing 3x the conductivity	2008	43
12	Poly(aryl ether sulfone) copolymer, PAES	60 mol% sulfonated PAES (sPAES-60)	Balance between the proton conductivity and MeOH crossover	2008	44
13	PI copolymer with phenyl benzimidazole,	Crosslinked	Acid impregnated high-*T* stable, efficient	2008	45
14	Poly(styrene-b-vinyl-benzyl phosphonic acid), PS-b-PVBPA	poly(2,6-dimethyl-1,4-phenylene ether), PPE	$T_g > 160°C$, thermal stability $\leq 350°C$; composition dependent	2008	46
15	Sulfonated PPE, SPPE	Bromo-methylated PPE, BrPPE aminated with n-propyl-amine, $PrNH_2$	Solution cast for DM-FC = 16 times better than Nafion	2008	47
16	Aromatic polyether bearing polar pyridine and phosphinoxide groups	Co-polysulfone for PEM strength and dimensional stability	Cast blends were doped with H_3PO_4 to 200 wt% gain, tested at 140°C	2008	48–50
17	Copolymer of 2,5-(di-hydroxy biphenyl)-pyridine and tetramethyl-biphenyl diol with bis(4-fluorophenyl) sulfone	Copolymers of 2,5-(di-hydroxy biphenyl)-pyridine tetra-methyl-biphenyl diol with phenyl phosphine oxide difluoride or decafluorobiphenyl	PEM were cast from blends of these copolymers. Conductivity at room-*T* of 10^{-3} increased to 2.5×10^{-2} S/cm at 180°C	2009	51
18	Poly(l,3-cyclohexadiene) PCHD	Diverse polymers or copolymers	PCHD may be aromatized. sulfonated or fluorinated	2009	53 53
19	Poly(benzimidazole), PBI	Poly(pyridine-co-sulfone) copolymer, PPycoPSF(1:1)	PEM immersed in 85% H_3PO_4 was used at 150°C	2009	54
20	Fluorinated sulfonated aromatic polyether	Modified PBI with $-S(O)_2-$ and $-C(CF_3)_2-$ linking groups	Good mechanical strength, proton conductivity, high-*T* performance	2010	55
21	Aromatic polyethers with pyridine groups	Copolymers doped with phosphonic acid	PEM with high modulus up to 230°C. $T_g \leq 260°C$ stability up to 400°C and conductivity $\approx 10^{-2}$ S/cm at 130°C		56

(*Continued*)

Table 3.2 Continued

No.	Polymer-1	Polymer-2	Comments	Year	Ref.
23	Soluble, containing sulfonic acid benzimidazole (SPBI)	Sulfonated poly(ether ether ketone) (SPEEK)	Dimensional stability in water at 80 °C). Proton conductivity and IEC = 55 mS/cm	2010	57
24	Poly(aryl ether ketone phthalazinone)	PAEK, PAK, PEEK, PEKK, PBI, thermoplastic PI, PEL, PAES, PPS poly or their mixtures	High-T miscible blends, with good processability. T_g = 108–365°C; T_m = 332–360°C	2011	58
25	Poly(2-acrylamido-2-methyl-1-propane-sulfonic acid) or poly(styrene sulfonic acid)	Poly(2.5-benz-imidazole (PBI)	At ratio 1:2 the proton conductivity = 0.1 S/cm	2011	59, 60
26	Polybenzimidazole, PBI and a miscible with it sulfonated polymer, e.g.. PS, PEEK, PAEK PSF, PPE, PPS	A modified PBI containing O and S; or a C-C bond, divalent aryl or aryl sulfone	PEM is free of water and exhibits a proton conductivity at T = 100 to 250°C of at least 10^{-5} S/cm	2011	61
27	Poly(vinylidene fluoride) (PVDF) and copolymers of PVDF	Poly[4(5)-vinyl-imidazole/vinylimidazolium trifluoro-methyl-sulfonyl-imide]	Strength controlled by PVDF (T_m = 160°C), conductivity by TFMSI content	2011	62

Notes: HFP, hexafluoropropylidene.

at $T > 150°C$ (for increased catalyst activity and reduced CO and other impurities), the PEM polymer must be thermally stable, resistant to oxidation, have good mechanical properties and high proton conductivity after doping with acid.[31–62]

3.4 Characterization of HTPBs for FC applications

Characterization methods may be segregated into three categories:

1. characterization of individual blend components;
2. characterization of polymer blends;
3. characterization for specific polymer blend application (namely, that of fuel or solar cell, of electrically conductive and used in aerospace).

There are a number of detailed descriptions of the fundamentals and characterization test procedures.[63,64]

3.4.1 Characterization of individual polymer components

Standard characterization methods are applicable to most polymers. However, as polymerization conditions are pushed toward high temperature and pressure, the side reactions may lead to contaminants on both sides – the low molecular weight fragments and partially crosslinked fraction. Thus, these three parts may require separate characterization.

Polymer conformation/configuration and interactions may be determined using the following standard methods.

- FT-IR spectroscopy over a wavelength range 550–4000 cm^{-1} with attenuated total reflection (ATR) module. Prior to measurements, the samples ought to be well dried (e.g., vacuum, $T = 80°C$). Thin, cast films or KBr pellets may be used. It is noteworthy that IR spectroscopy detects functional or polar bonds, e.g., C=O, O-H, etc.
- Since Raman spectroscopy sees carbon allotropes and polarizable carbon bonds (e.g., C-C, C=C, etc.) it offers complementary information to FT-IR. The Raman wavelengths of interest are 500–2000 cm^{-1}. For example, membrane interaction with H_3PO_4 was analyzed following the shift of the imidazole and pyridine groups at 1539 and 1592cm^{-1}, respectively. The spectra were recorded using the near-infrared spectroscopy (NIR) excitation at 1064 nm.[54]
- The solution ^{1}H and ^{13}C nuclear magnetic resonance (NMR) spectra are recorded at frequencies of 250 and 62.9 MHz, respectively, with tetra-methyl silane, tri-methylsilyl (TMS), as internal reference. Before

analysis, sulfonated compounds are to be converted to the sodium sulfonate form.[65] The solid state NMR spectroscopy with the magic angle spinning at frequency 1–35 kHz, magic angle spinning NMR (MAS NMR), is often performed on dilute or rare nuclei (e.g., ^{13}C, ^{15}N, ^{29}Si, ^{31}P) at an appropriate Larmor frequency $\omega = 200$–850 MHz (the Larmor frequency is a product of the magnetic field strength, B_o, and the characteristic for each nuclei gyro-magnetic ratio, γ: $\omega = \gamma B_o$). Spectra may be referenced to TMS (^{1}H) or phosphor (^{31}P). High resolution 1H MAS NMR provides information on hydrogen bonding and proton mobility. Since the latter on the one hand determines the proton conduction and on the other in solid state polyelectrolyte's proton motion is connected with chain dynamics, knowledge of the latter is essential for understanding the PEM transport mechanism.

- Size exclusion chromatography (SEC) requires dissolution of polymers in a good solvent at T that ascertains full miscibility. The system consists of a pump, a multi-angle laser light scattering detector (with 18 detection angles), a refractive index detector and an in-line viscometer. The flow rate is about 1.0 mL/min. The SEC columns are calibrated with a mono-dispersed polymer (e.g., polystyrene (PS)) of well-known molar mass, plotting molecular weight (MW) vs. elution volume. Comparing the test results to the calibration data provides full information about MW and molecular weight distribution (MWD). However, since the standard SEC operates at $T \leq 135°C$ it may not be useful for some HTP. The alternatives are the high temperature scattering and solution viscosity measurements. In the latter case, for better precision the intrinsic viscosity, $[\eta]$, should be determined using a thermostated viscometer.[66] Alternatively, for comparisons, the inherent viscosity, $\eta_{inh} = [\ln(\eta/\eta_0)]/c$, may be determined using a single solution at a single temperature (e.g., 0.5 g/dL solutions in 96% H_2SO_4 at 30°C).
- Thermogravimetric analysis (TGA) may be carried in air or under N_2, either as a dynamic temperature scan (e.g., 40–700 °C at a rate of 10 K/min) or as an isothermal performed at constant temperature vs. time (e.g., at $T = 350°C$, for $t = 90$ min); the conditions depend on the tested polymer or blend.
- Differential scanning calorimetry (DSC) is being used for studying the vitrification, T_g, melting, T_m, and crystallization, T_c, transitions. The measurements are usually performed under N_2 at $T = -80$ to 340°C at a heating rate of $q = 5$–20 K/min in cycles consisting of 1st heating-cooling 2nd heating scans.
- The pyrolysis-gas chromatography-mass spectrometry, GC-MS, measurements are performed using a pyrolyzer and gas chromatograph coupled to mass-selective detector. The ~100 mg sample is placed in a quartz sample tube, preheated at 140°C to remove the volatiles and then rapidly heated to the selected pyrolysis temperature, $T_{pyr} = 600$–$1000°C$.

The volatile fragments are separated by gas chromatography (GC) and identified by MS.

- Melt rheology offers a spectrum of information about the polymer (or blend), its thermal stability, shear sensitivity, the mechanism of degradation (e.g., chain scission vs. branching and crosslinking), processability, etc. The preferred method of testing is dynamic between cone-and-plate or parallel plates.[67]

3.4.2 Testing polymer blends

The following standard methods may be used.

- For polymer blends the key is the knowledge of miscibility.[68–70] Since miscibility and its type depend on many variables, the most desirable thing is to generate a phase diagram reflecting the range of variables within which the blend will be prepared, processed and used, for example all components of the blend including compatibilizers, stabilizers, additives, concentration, temperature, pressure, etc. As an alternative, one may determine the miscibility of a single composition under a single set of conditions. In this case, the residual stresses, time scale for their dissipation and invariable blend coarsening must be taken into account. Blends have three phases: miscible, immiscible and the metastable meso-phase between them (excepting the critical point). The domain size in miscible system: $d \leq 2$ nm.
- For the use of HTPBs in FCs the films morphology is highly pertinent. The morphology depends on the thermodynamic interactions (the phase diagram), the stress field during forming, the coalescence/coarsening time and conditions, as well as the post-treatment (e.g., extraction of one phase). Depending on the domain size, different microscopic methods are used, namely optical (OM), SEM, AFM or transmission electron microscopy (TEM).
- Owing to the wide range of magnification, $\times = 10$–500,000, the most commonly used in FC research and technology is SEM.[71] To avoid electrostatic charge at the surface, specimens should be coated with, for example, Au/Pt layer and electrically grounded. Coating is done by low vacuum sputtering, high vacuum evaporation or by solution coating. In some cases it may be necessary to improve Au/polymer adhesion by treatment with di-thiol solution. Recently new SEMs labeled as environmental, low-vacuum, high-pressure, or variable-pressure were developed. The higher pressure minimizes outgassing from volatile samples and diminishes charging the nonconductive samples.

3.4.3 Testing polymer blends for specific applications

The following standard methods may be used.[72]

- The titration of acid functions of the polymeric membrane is performed after immersing it for 24 h in a 2M NaCl aqueous solution. The mixture is titrated with a 0.05M NaOH aq. solution to pH = 7.0. To test an ionic polymer for its alkalinity the blend is titrated with HCl.
- The gas permeability through the hollow fiber single cell is performed by placing the test cell in a pressure vessel at a controlled temperature. The permeation rates of pertinent gases (e.g., He, CO_2, O_2, N_2) are measured at ambient pressure and temperature using a soap opti-digital flowmeter. The pressure difference across the hollow fiber should be kept constant at 0.5 MPa. The permeability is calculated using the relation $p_i = Q_i/[75n\pi d_{out}\Delta P]$, where Q_i (mL/s) is the flow rate of gas i, n is the number of hollow fibers in the cell, d_{out} (cm) is the fiber outer diameter, L (cm) is the fiber length, ΔP (MPa) is the pressure difference across the membrane. The separation factor relative to N_2 is written as $\alpha_i = p_i / p_{N2}$.[73,74] For binary gas pairs O_2/N_2, CO_2/CH_4, H_2/N_2, etc., the gas separation by polymeric membranes shows a tradeoff between permeability and selectivity. According to Robeson, empirically there is a relation between the gas separation factor and permeability of the more permeable gas $\log(\alpha_{A/B}) = a_0 - a_1 \log(p_A)$, where a_i are empirical parameters for the given pair.[75–77]
- Flat, 70 μm thick membranes obtained from a solution of PBI with poly(pyridine-*co*-sulfone), were immersed in 85% H_3PO_4 until gaining 180–200 wt%. The membrane (surface area = 960 mm^2) was mounted in a test cell equipped with a quadrupole mass spectrometer for monitoring of the outlet. Different gases were flown through the anode chamber of the cell at a rate of 80 mL/min, while Ar was flown through the cathode chamber at 10 mL/min. Outlets of these chambers were connected to a quadrupole mass spectrometer, so the gas composition from both outlets was recorded. Testing provided information on the amount of gas that permeated at T=130–170°C through the HTPB membrane, separating anode from cathode chamber.[54,78]
- Methanol (MeOH) permeability at 30°C was determined following the electrochemical monitoring method. Accordingly, the membrane was clamped between the two compartments of the test cell, both magnetically stirred. The required volume of aqueous MeOH and KOH solution was added to compartment 'A' containing a Pt-mesh counter electrode (CE). An equal volume of KOH aqueous solution was added to compartment 'B', containing a smooth Pt-working and Hg/HgO reference electrodes. The cell potential was recorded every 30 min at a

rate of 50 mV/s between 0.2 and 1 V. The height of the MeOH oxidation peak was assumed to be proportional to the amount of MeOH permeated through the membrane. Gas chromatographic analysis was also carried out.[79]

- The MeOH permeability was determined by calculating its flux through the membrane placed into a glass cell in a double bath at constant temperature.[44] One bath was filled with 5 wt% MeOH aqueous solution, and the other with water. The MeOH concentration was calculated from the refractive index. Sampling was done for 5 h every 10 min. As a first approximation the MeOH permeability is proportional to the rate of concentration change.
- Membrane water uptake (WU) may be defined as the amount of water absorbed by it while the specimen is stored in a container at constant temperature and relative humidity (RH) until equilibrium. The type of bonding responsible for the membrane hygroscopicity may be determined by ^{1}H-MAS-NMR. The weight-based water sorption is expressed as

$$WU_W\ (\%) = 100(w_{wet}/w_{dry} - 1). \qquad [3.1]$$

- The ultimate WU may be determined for membranes from which excess sulfuric or phosphoric acid is removed by storing it in H_2O for 24 h. Next, WU is measured by immersing several PEM pieces (around 0.1 g) in deionized water at 20°C or 25°C for 24 h. Next, each specimen is wiped with tissue paper and weighed. The water uptake is calculated from Equation [3.1] as the average weight ratio of absorbed water on the dry sample weight. The dimensional stability should also be calculated. The amount of free and bound water in PEM may be determined in DSC by scanning the specimens from −50 to +50°C at 1°C/min. The amount of free water is calculated from the area under the melting peak at 0°C and the heat of fusion $\Delta H(H_2O) = 334$ J/g. Bound water is expressed as the difference between WU_W and the free water content. The PEM water uptake may also be determined using TGA, heating the specimen from −30°C to 900°C at a rate $q = 10$°C/min.[59,60]
- The PEM ion-exchange capacity (IEC) is defined as the number of milli-equivalents (meq) of acidic ions per 1 g of the dry polymer. IEC is determined by titration. First, dry PEM specimen (around 0.1 g) in an acid-form is immersed in 50 mL 2M aq. NaCl or KCl for 48 h. Next, three 10 mL aliquots are removed and titrated with 0.01 M NaOH to a methyl red endpoint. The IEC is calculated as an average of three tests. Alternatively, instead of NaOH a very dilute aq. Na_2CO_3 solution may be used. Good PEM has IEC ≈ 1 meq/g and the equilibrium water swelling B ≥ 119%, both calculated on PEM dry weight.

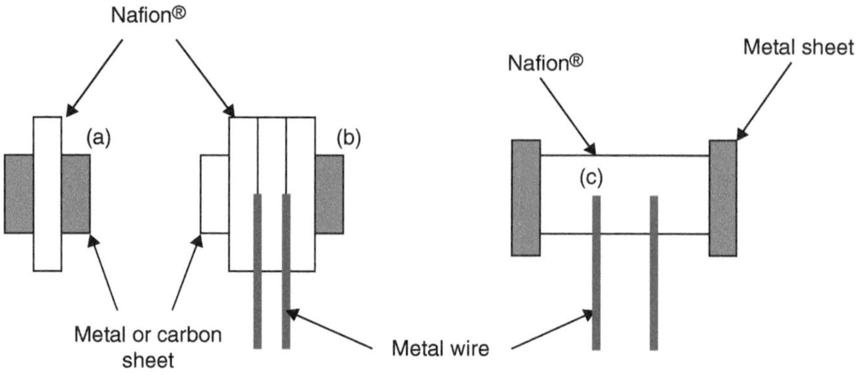

3.2 Arrangements for membrane proton conductivity tests in the thickness direction with 2-probe (a) and 4-probe (b), as well as 4-probe in the surface direction (c).[81]

- Proton conductivities of fully hydrated membranes (24 h at ambient temperature in double deionized H_2O) may be measured using two- or four-probe electrochemical impedance spectroscopy (EIS) at frequency 0.1–10 MHz with AC amplitude of 5 or 10 mV (Fig. 3.2).[44,80,81] For good membrane-electrode contact the PEM is placed between two Hg or Pt electrodes in a sealed conductivity cell, thermostated at the desired T for about 5 h before measurements. It is advisable to perform the measurements with dry membranes from 20 up to 100°C in 10°C steps with wet membranes. Each sample should be measured 10 times and the average value of the impedance, R, used for calculating the proton conductivity: $\sigma = d/RS$ (S/cm), where d is the membrane thickness, thus the distance between the electrodes. The results are sensitive to the specimen immersion depth, quality of deionized water, and electrode/membrane contact. Usually, the ionic conductivity correlates with the degree of sulfonation.[38,64,82,83]
- Proton conductivity is proportional to the number of charges (n) per unit volume and the ion mobility in an electric field (μ): $\sigma = n\mu$. The proton conductivity was measured by dielectric spectroscopy in a two-electrode geometry using an SI 1260 impedance/gain-phase analyzer in the frequency range of 10^{-1}–10^{-6} Hz. The humidity of air during data acquisition was set by mixing dry nitrogen with humidity saturated nitrogen. The relative humidity (RH) was measured using a Sensiron SHT15 sensor. For dry-condition proton conductivity measurements the samples were dried at 50°C under vacuum for four days prior to the test. Completely dry N_2 was used to flush the samples during the measurements. The samples for conductivity measurements were pressed to tablets and contacted by E-tek® and stainless steel electrodes. Proton conductivities were

evaluated from both Cole-Cole and Bode plots. All curves have a specific frequency region in which AC conductivity is frequency independent (a plateau). The extrapolation of the plateau to $\omega = 0$ provides DC conductivity. The precision of this method was checked by comparison of the results with values obtained via Cole-Cole plot. The data obtained by these approaches were in good agreement.

- The temperature dependence of the conductivity in polymer electrolytes (σ) may indicate the conduction mechanism, for example the Arrhenius type: $\ln\sigma = \ln\sigma_o - E_a/RT$ (E_a is the activation energy, R is the gas constant, and σ_o is a pre-exponential factor).[84] In systems that obey the Arrhenius dependence the 'thermally activated' ions have sufficient energy to overcome the E_a energy barrier, i.e., thus this behavior is mainly observed at high T.[85] At lower temperatures the Vogel-Fulcher-Tamman-Hesse (VFTH) equation applies: $\ln\sigma = A_o + A_1/(T - T_o)$ (where A_o is a pre-exponential factor, A_1 is a constant, T is the temperature of measurement and $T_o \approx T_g - 50$ is the reference temperature). However, in the case of proton conduction, T-dependence is complicated by the temperature-dependent dissociation that generates 'activated' protons able to overcome the energy barriers for conduction. Experimentally, the PEM conductivity when scanned from 20°C to 150°C was higher than when scanning in the opposite direction from 150°C to 20°C, possibly because of the reduction of residual water and self-condensation of the phosphonic acid groups. The proton conductivities of dry PEM followed the Arrhenius equation. However, in a fuel cell PEM contains a variable quantity of water produced at the cathode; thus membrane conductivities at different RH should be known. In general, at constant temperature the proton conductivity follows the relation: $\ln\sigma \cong B_0 + B_1(RH)$, where B_i are parameters. The data plotted as σ vs. WU (not RH!) reach a plateau at high water content.
- Thermal analyses (e.g., TGA, differential thermal analysis (DTA), DSC) of the HTPB membranes follow the same procedure as that for HTP, discussed earlier. Evidently, the presence of two polymers may be affected by trans-reaction, by the unexpected effects of one set of additives on the other polymer set, etc. Also, at higher concentration of acidic groups they may convert into anhydrides.
- The accelerated membrane degradation tests include: Fenton test, thermal degradation test, potential load cycling, and relative humidity cycling. The Fenton's test determines PEM chemical resistance. Thus, 20 ppm $FeSO_4$ is dissolved in H_2O_2, the membranes are immersed in the solution at 80°C for 6 h. Next, the PEM is washed with H_2O, dried under vacuum at 60°C for 3 h, and weighed. The PEM resistance to oxidation is judged by the absence of weight loss and increased brittleness.[86]
- Morphology is of key importance for the performance of polymer alloys and blends (PAB). Diverse microscopic and spectroscopic

methods have been used. **Optical microscopy (OM)** and polarized optical microscopy images are important for highly polydispersed morphologies. These should be measured using a heating plate and environmental chamber with dry N_2. **Scanning electron microscopy (SEM)** and field emission SEM are versatile sources of micrographs in a wide range of magnification. For example, SEM offers informative images of the PEM surface and cross sections (sputtering with Au/Pd may be required). **The transmission electron microscopy (TEM)** images require labor-intensive specimen preparation, including preparation of carbon replicas, but it may be the best choice for high resolution, nanoscale imaging. **The atomic force microscopy (AFM)** expands its applicability from microscopy to the nanoscale mechanical and viscoelastic tester. **The acoustic microscopy (AM)** has been known for over 30 years and it found new application in polymer blends. AM uses phase-contrast ultrasonics (e.g., 1.2 GHz frequency) for detecting variation of polymer rigidity (e.g., the presence of voids between polypropylene (PP) spherulites); thus in principle it may provide direct information on the strength of the interphase between the polymeric components.[87]

3.5 Solar cell (SC) applications

Solar cells (SCs) or photovoltaic (PV) devices convert sunlight into electricity. A. Edmond Becquerel is credited with the discovery of the PV effect in 1839. However, the phenomenon remained of little practical interest until 1954, when a photocell based on high purity crystalline silicon *p–n* junction increased the efficiency of solar energy-to-electricity conversion to 4.5–6.0%.[88] Figure 3.3 displays the historical increases of PV efficiency based on inorganic or organic material. The data are for different constructions and materials, for example the inorganic cells are based on Si, Ge, GaInP or GaInAs. The increase of efficiency is spectacular – from 0.5% in 1942 to 40.8 % in 2008, and 42.43% in 2010.[89–91] The organic PVs comprise dye-sensitized,[92,93] low molecular weight or polymeric organics as well as inorganic/organic hybrids.[94] Green *et al.* published lists of the independently confirmed efficiencies for SCs separated into individual categories.[95]

There are fundamental differences between the inorganic SC and the organic SC (OSC). In the former, the charge carriers (electrons and holes) are not tightly bound to each other, and once separated from each other by electric field they travel to the respective electrodes. In OSCs the electrons and holes form strongly bound excitons, which dissociate at electrode interfaces or within the interphase between organic donor and acceptor. Furthermore, in OSCs the majority of carriers exist in different phases and

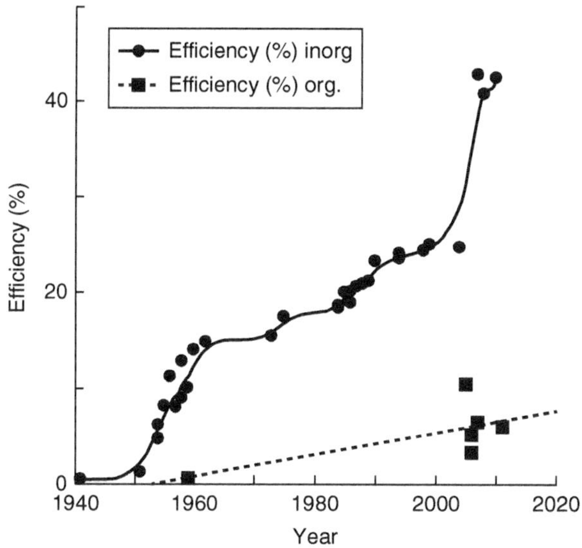

3.3 Efficiency of diverse types of inorganic or organic solar cells.

their movements create the current.[96] The difference in the mechanism of charge generation is the fundamental reason for the difference in their level of efficiency, dramatically evident from the data in Fig. 3.3.

Polymeric solar cells (PSCs) are a type of OSC. They are being intensely studied as alternatives to inorganic SCs, replacing expensive raw materials and the costly, high temperature and high vacuum manufacturing technology. Considering the well-established large-scale polymer processability, it is expected that PSCs will be more economic, especially on large-scale production. Furthermore, PSCs offer toughness, pliability, flexibility and light weight. These advantages over the crystalline systems counterbalance the low efficiency and photo-degradability, the latter solvable by proper SC design and stabilization.[97] Evidently, the needs for energy production in space and on the ground, for large-scale electricity production and supplying of electronic devices, are quite different and it is most likely that the current development will end with solar cells of different types being optimized for different needs.

From the view of construction, there are four types of photovoltaic solar cells:

1. single layer;
2. bi-layer;
3. bulk heterojunction (BHJ);
4. tandem or multi-junction.

Single layer cells are those where a 40 to 200 nm thick layer of organic electronic material is deposited between the two metallic conductors, in which transparent indium tin oxide (ITO) is the hole collector and a layer of Al, Mg or Ca is an electron collector. For example, poly(9,9′-dioctylfluorene-*co*-benzothiadiazole) (F8BT), poly(9,9′-dioctylfluorene-*co*-bis-*N,N*′-(4-butylphenyl)-bis-*N,N*′-phenyl-1,4-phenylenediamine) (PFB), and their blends were used as PV with the power conversion efficiency, PCE = 0.1–0.5%.[98]

Bi-layer cells are those with the two photo active semiconducting components placed one over another, for example a conjugated polymer as electron donor (polyphenylenevinylene) and an acceptor (red emitting polyfluorene) with PCE = 1.5%.[99,100]

Bulk heterojunction (BHJ) cells have blended donor and acceptor forming large interphase. These solar cells are mostly made of poly-conjugated polymer (donor) and a fullerene derivative (acceptor). They have several advantages over single- and bi-layered cells, namely a thick active layer, a large interphase area created by an interpenetrating and co-continuous morphology, as well as relative ease of processing. All these translate into improved efficiency. Thus, the multi-layered structure is composed of ITO/the anode buffer blend of poly(3,4-alkenedioxy-thiophenes):poly(styrene sulfonate) (PEDOT:PSS)/the active blend layer of poly[2-methyl, 5-(3′,7′-dimethyloctyloxy dimethyl-octyloxy)-*p*-phenylene vinylene]: [(6,6′)-phenyl C_{61}-butyric acid methyl ester] (PMDMOPPV:PC_{60} BM) and LiF/AL gave PCE = 2.5%.[142] (see Figs 3.4 and 3.5).

The phase segregation in BHJ systems is of critical importance, but difficult to determine (nm scale and a gradient in the thickness direction). The best method is the use of AFM and TEM on cross-sectioned specimens. Recently Venkataraman *et al.* discussed a variety of molecular architectures for the packing of π-conjugated organic molecules to structures relevant for PV devices. The aim has been to optimize and balance the charge transfer interactions and packing interactions in PV through nanoscale segregated morphologies for efficient charge separation and transport.[101]

Creation of reproducible morphology of immiscible blends is notoriously difficult. The task is simpler when the system comprises a block copolymer

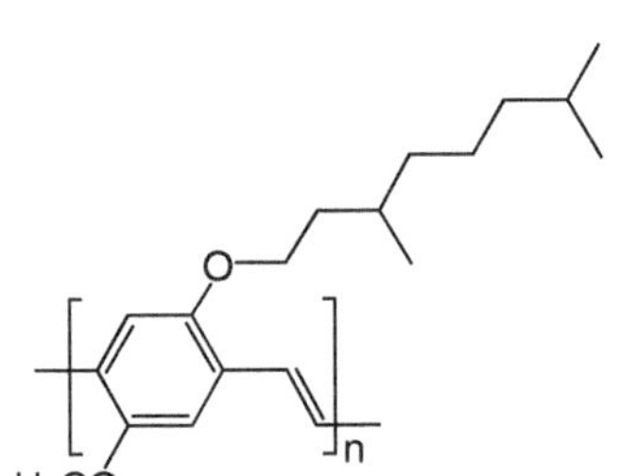

3.4 One component of active blend layer.

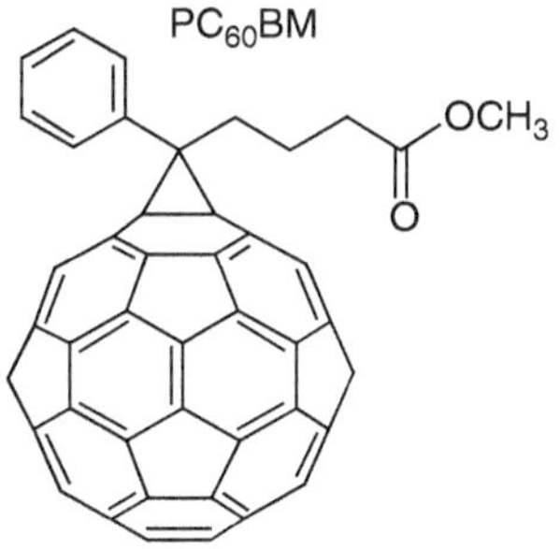

3.5 Additional active layer material in addition to Fig. 3.4 polymer.

(BCP).[69] Accordingly, instead of blending a conjugated polymer with fullerene derivative, one may start with such a polymer having grafted fullerene side groups. Now, the equilibrium morphology depends on the copolymer composition and, if reproducibly cast and annealed, it should form reproducible morphology. Recently Topham *et al.* reviewed the application methods of block copolymers in photovoltaic devices.[102] The copolymers are expected to overcome the two principal weaknesses of OSCs, namely the lack of long-term stability and poor efficiency. Considering the multi-parameter influences on efficiency of BCP devices, optimization may be long and difficult (the current maximum value of BCE-based OSCs is PCE = 1.7%).

Other solutions may also lead to self-assembly of well-designed morphologies. For example, building liquid crystals with attractive-repulsive oligo-chain ends creates such a possibility.[103] Park *et al.* used not synthetic, but an engineering approach to the problem by controlling '**e**vaporation of **s**olvent through a **s**urface of **enc**apsulating and **i**nduced **al**ignment (ESSENCIAL)'. The key is the use of a vapor-permeable cover layer, which allows for solvent evaporation while protecting the PV surface, but still allowing for the shear flow under application of pressure. The process (applicable to a roll-to-roll processing) leads to an optimized morphology, more uniform distribution of components and control of crystallinity.[104]

Getachew *et al.* recently published an interesting new BHJ composition. [105] The authors reported that PSC blends perform better when a fullerene-derivative ($PC_{60}BM$ acceptor) is blended with a 1:1 mixture of two thiophene-based polymers, D1 and D2, having the same conjugated backbone but different types and volume fraction of alkoxy side chains on the phenylene-ethynylene unit (Fig. 3.6).

The blend enhanced charge carrier mobility, which improved the ternary blend photovoltaic performance in comparison to binary blends of D1 or D2 with $PC_{60}BM$. The AFM showed that the improvement is related to finer morphology in the ternary system of the active layer.

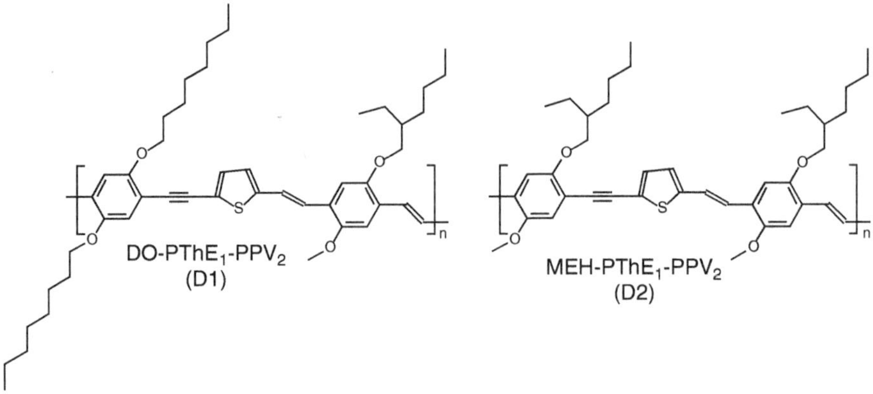

3.6 Donor polymers: D1 = poly{1,4-(2,5-dioctyloxy)-phenyleneethynylene-2,5-thiophenylene-vinylene-1,4-[(5-(2-ethylhexyl)oxy)-2-methyloxy]-phenylene-vinylene} and D2 = poly{1,4-(5-[(2-ethylhexyl)-oxy]-2-methoxy)-phenylene-ethynylene-2,5-thiophenylene-vinylene-1,4-[(5-(2-ethylhexyl)oxy)-2-methyloxy]-phenylene-vinylene}.

Kimber *et al.* carried out the first simulation of *bi*-continuous structures (e.g., disordered blend, gyroid and vertical rods) for PSCs. The morphology influences device performance as the interface between the electron and hole transporting phases divides the volume into two disjoint sub-volumes. The simulated *I–V* characteristics that disordered blends have lower performance than gyroids, but the morphology of vertical rods is the best.[106]

Tandem or multi-junction cells are built with two or more subcells (absorbing in different wavelength) arranged in stacks. From the point of view of electrical performance they may be considered connected in series, with the voltage given by the sum of subcell voltages and the current limited by the minimal of the constituent subcells, i.e. open circuit voltage: $V_{oc} = \sum_{i=1}^{n} V_{oci}$ and short circuit current: $I_{sc} = \min(I_{sci})$. The invention of multi-junction structures was responsible for the spectacular jump of efficiency in Fig. 3.3. For example, working with the US National Renewable Energy Laboratory (NREL), Spire Semiconductor announced in October 2010 a record PCE = 42.43% for a triple-junction metamorphic solar cell. The tandem stack had a mismatched InGaAs cell epitaxially grown on the back-side of a lightly doped, *n*-type GaAs wafer, which was flipped, before GaAs and InGaP cells were grown on the opposite wafer surface.[107] The psychological barrier of 50% efficiency is projected to fall within the next 10 years, assuming advancement of the current cell technology and sunlight concentrators.[108]

By contrast with the crystalline semiconductors, the efficiency of tandem OSCs is much lower. For example, tandem cells made from a combination of a poly(3-hexylthiophene-2,5-diyl):[6,6]-phenyl-C_{61}-butyric acid methyl ester and a Cu-phthalocyanine:fullerene subcells gave PCE = 4.6%, i.e., approximately a sum of efficiencies of individual subcells.[109]

The OSCs are built using electron donor and acceptor PV or thin-film *p-n* semiconductors, which may be deposited on other polymers for solar cell applications.[110,111] In most PSCs the polymers are conjugated with delocalized π-electrons, which makes them conductive, and for the PV applications they absorb light delocalizing the molecular orbitals. The process depends on the wavelength π–π* transition. In most PSCs the anode (transparent in full range of visible light) is the brittle and expensive indium tin oxide (ITO), which along with the glass substrate may be replaced by, e.g., poly(3,4-ethylene-*di*-oxythiophene)/poly(styrene-sulfonate) (PEDOT/PSS). The efficiencies of these cells on glass and plastic substrates are comparable to those of ITO-based devices, but in bending test ITO breaks after a few cycles while the all-polymer cell shows no effect of bending after hundreds of cycles.[112]

Currently (see Fig. 3.3) the efficiency of the best organic SC is about 10%. The evolution of the PV technology may be divided into three 'generations':[113]

- The first generation was based upon thick Si-wafers, which could be etched, texturing the surface and enhancing absorption of photons. Efficiency of the laboratory-scale devices approached the theoretical limit for single junction PV of 31%.
- The second generation focused on the low cost thin film devices with low efficiency, inorganic or organic PV. The former were based on thin-films of amorphous, microcrystalline or polycrystalline Si, CdTe, and copper-indium-gallium *di*-selenide [$Cu(In,Ga)Se_2$, or CIGS]. Organic PVs include those based on dyes, liquid crystals, conjugated polymers, organic glasses, etc., which show *p*-type or *n*-type semiconducting properties.
- The third generation with shaped nano-particles and quantum dots (QD) boosts the efficiency (so far only in inorganic SCs) by about 80% and may have a similar effect in PSCs. The attempt is preventing energy loss as heat (thermalization) of the photoexcited electron–hole pair and their loss (caused by the transparency to photons with energy less than the semiconductor band gap). The two effects limit the theoretical conversion efficiency of a solar cell to about 44%; without them the limit would be around 94%. Utilization of the quantum dot approach may provide a significant improvement of PV efficiency and thus better economy.[114] Another approach taken at the Oak Ridge National Laboratory is the formation of *n*-type ZnO nanocones penetrating by around 100 nm into the *p*-type poly-crystalline (PX) CdTe matrix (absorber and hole transporter) on transparent conductive ZnO (electric field generator and

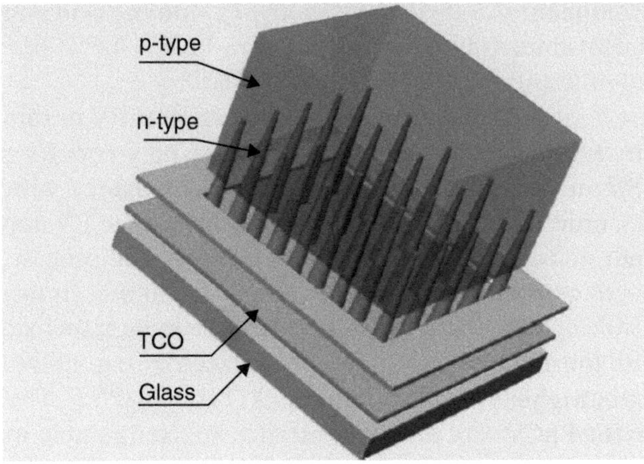

3.7 *n*-type ZnO nanocones penetrating the *p*-type PX CdTe matrix on transparent conductive ZnO (TCO) and glass substrate.

electron transporter) and glass substrate (see Fig. 3.7).[115] The nanocone type SC has 3.2% efficiency (without optimization), while the single-layer, planar ZnO-PX CdTe cell has PCE = 1.8%. This circa 80% higher SC efficiency was achieved in spite of smaller area and reduced light absorption.

Since the 1950s, organic solar cells (OSCs) have been constructed using organic dyes and molecules with semiconducting properties (e.g., methylene blue, anthracene), biological molecules active in photosynthesis (e.g., chlorophylls, carotenes) and then in the 1980s, the conjugated, conductive polymers [e.g., polyacetylene (PACE), polyaniline (PANI), polysulfur nitride $(SN)_x$]. The conversion efficiencies of around 0.1% increased ten-fold when *bi*-layered design was introduced, and then by additional factor of five by blending the electron donor with the acceptor, creating the bulk heterojunction. By the year 2008 the power conversion efficiency of dye-sensitized solar cells (DSSCs) under standard test conditions was PCE = 11.3%.[116]

Several books and reviews on organic and polymeric solar cells (OSCs and PSCs) are available.[16,117–122] The use of polymers for PV applications leads to longer shelf/operating lifetimes and larger dielectric constants than other organic materials. The use of conjugated polymers as a *p*-type semiconductor and a derivative of the fullerene C60 as *n*-type material dominates *bi*-layer and BHJ devices with PCE = 7.4 and 8.3%.[123,124] A recent patent describes an organic light emitting diode-PSC device in the form of a multi-layered sandwich.[125] The BHJ-type polymer layers are 50 to 200 nm thick and made of, for example, poly(3-hexylthiophene)/[6,6]-phenyl C_{61}-butyric acid methyl ester (P3HT/PCBM). Other polymers may be used as the donors, for example, poly[2-methoxy-5-(3,7-dimethyloctyloxy)-1,4-phenylene vinylene] (PMDMO-PPV), or poly(2-methoxy-5-(2′-ethyl-hexyloxy)-1,4-phenylene vinylene) (PMEH-PPV). Besides $PC_{60}BM$, [6,6]-phenyl C_{71}-butyric acid methyl ester ($PC_{70}BM$), or single-walled carbon nanotubes (SWNT) may be used as the acceptor. The two new derivatives of C_{70}-fullerene: (1) phenyl-C_{71}-propionic acid propyl ester ($PC_{70}PP$) and (2) phenyl-C_{71}-propionic acid butyl ester ($PC_{70}PB$), are expected to increase efficiency of this type of PSC. Their blends with PMDMOPPV or P3HT as a donor polymer showed better miscibility and performance (e.g., 10% higher short circuit current density).[126] Recent articles describe efficient PSCs based on modified P3HT or a new broad band gap donor-acceptor conjugated copolymer with 2-alkylthiophene-substituted benzodithiophene (BDT) donor segment and a thiazolo[5,4-d]thiazole (TTZ) acceptor unit (PBDTTTZ).[127,128]

For the PSC to become competitive with the inorganic SC their PCE should increase to over 10% and lifetime to over 1 year.[16] The great advantages of OSCs or PSCs are their weight, flexibility and ease of processing, either from solution or film extrusion and printing. Because of the desired conjugation, planarity of the polymeric chains and slenderness of their layers, they are readily attacked by O_2 and humidity. Some years back, when these problems started affecting longevity and performance of the conducting 'polymer wires', the solution was a co-extrusion coating with a barrier polymer. This method, as well as an increase of the polymer oxidation potential, is being explored for PSCs. Furthermore, there is a continuous development of the electron donor and acceptor materials (including the self-assembly systems) that engender specific and stable morphology.[129] In 2008 Konarka Technologies started production of polymer-fullerene SCs with the GW/y capacity, initially for laptop-recharging briefcases, tents, umbrellas, and as semi-transparent window tinting; the future for PSCs looks bright (no pun intended).

3.6 Characterization of HTPBs for polymeric solar cells (PSCs)

As was the case for the fuel cells, the characterization methods for PSCs may be divided into the three categories:

1. individual polymers;
2. their blends;
3. the device, i.e. the PSC.

Evidently, the methods for characterization of polymers and their blends do not vary much with their intended applications, excepting tests for specific critical properties for the envisioned purposes. In the case of PSCs the most important is the ability to convert light into electricity and vice versa, which very much depends on the construction and molecular as well as domain morphologies. Thus, in this section the characterization of SCs will be outlined.

The process of conversion of light into electricity by an OSC follows these steps.

- Absorption of a photon by active layer creating excitons, i.e., electrically-neutral electron-hole pairs.
- Diffusion of excitons to the donor/acceptor interface where they separate into charge carriers: electrons excited to the lowest unoccupied molecular orbital (LUMO) leaving holes in the highest occupied molecular orbital (HOMO). The HOMO-to-LUMO energy difference is known as the band gap < 4.8 eV.[130]
- Transport of carriers to the respective electrodes.

The SC is characterized by its current–voltage dependence measured under standard solar simulator (CEI/IEC 904-3) as global air mass 1.5 (AM 1.5G) with the power density of 1000 W/m^2 corresponding to spectral intensity distribution for wavelength 250–2500 nm and at an incident angle of 48.2° (Fig. 3.8).[131] The intensity of the illumination should be verified before the measurements using a calibrated silicon diode with known spectral response.

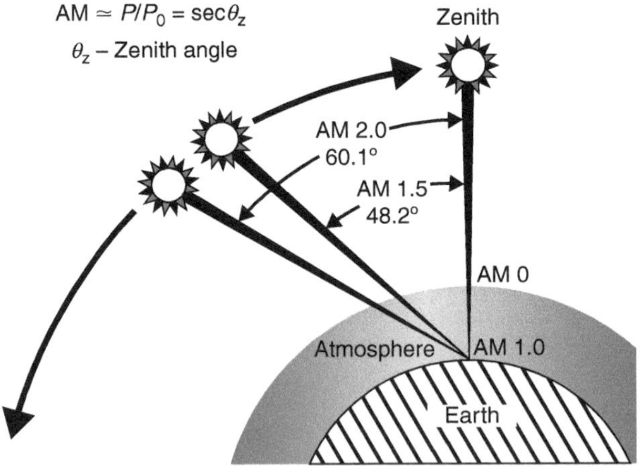

3.8 Method of characterizing SC current–voltage dependence.

Several PV parameters should be defined.[132]

- Open circuit voltage (V_{oc}) – The maximum voltage across a PV cell in sunlight when no current is flowing.
- Short circuit current (I_{sc}) – The current that flows through an illuminated solar cell when there is no external resistance, thus when the electrodes are short-circuited.
- Power (P_{in}, P_e) – The incidental power of solar radiation, and electrical power generated by the PV cell, respectively.
- Maximum power point (I_{mpp}, V_{mpp}) – The point on the I–V curve where the maximum power is produced: $P_e = \lim_{P_e \to \max} I \cdot V = I_{mpp} \cdot V_{mpp}$ (see Fig. 3.9).
- Fill factor (FF) – The ratio of a photovoltaic cell's actual maximum power output to its theoretical power output. FF depends on the fraction of charge carriers reaching the electrodes compared to those that recombined, thus on $\tau\mu$, where τ is the lifetime and μ the mobility.
- Power conversion efficiency (PCE or η_{PCE}) – The ratio of power output to power input.
- External quantum efficiency (EQE) – The percentage of photons hitting the photoreactive surface that produce an electron–hole pair in photosensitive devices – a measure of the device's electrical sensitivity to light, thus:

$$\mathrm{EQE}(\lambda) = 100 \times \left[I_{sc}(\lambda)/e\right] / \left[\mathrm{P}_{in}(\lambda) e\lambda\right] \qquad [3.2]$$

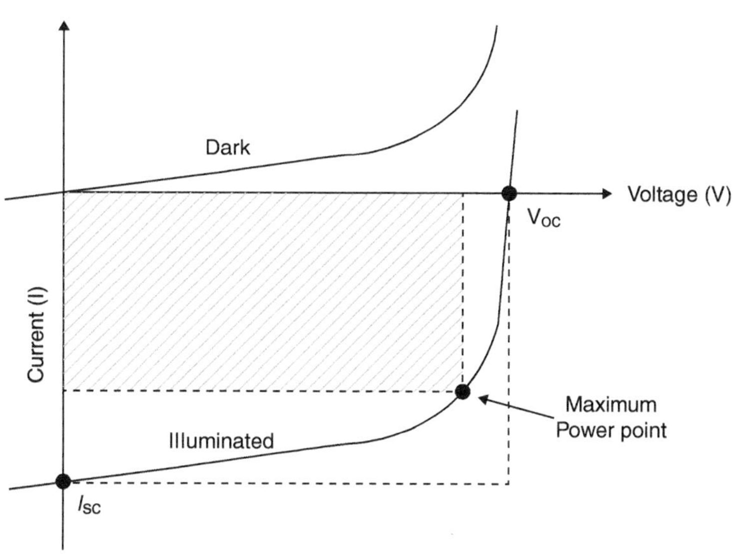

3.9 Determination of maximum power output from current–voltage plot.

with $e = 4.80320427(12)\times10^{-10}$ *statcoulomb* being the elementary charge.

Depending on the PV design, location of the EQE maximum ranges from $\lambda = 400$ to 1000 nm.

These quantities are interrelated, namely

$$P_e = I_{mpp}V_{mpp} = \eta_{PCE}P_{in} = I_{sc}V_{oc}FF \quad [3.3]$$

For accurate PV measurements one uses a solar simulator with a light spectrum similar to that of AM 1.5G and a calibrated reference cell to set the intensity. The SC measurement is divided into two steps: (a) the determination of the solar simulator spectral mismatch factor, *M*; and (b) measurement of the *I–V* curve of the solar cell referenced to standard test conditions (STC). The match between the simulator and the AM1.5G reference spectra [$E_S(\lambda)$ and $E_R(\lambda)$, respectively] is not perfect. Furthermore, while for crystalline Si SCs, $M = 1.00 \pm 0.02$ (stable calibrated reference cells are available), for OSCs suitable and stable reference cells do not exist – in this case $M = 0.9$ was reported.[133]

The maximum available current and voltage (I_{mpp}, V_{mpp}) of a *p-n* junction device is given by the quasi Fermi levels of *n*-doped and *p*-doped semiconductor energy levels. In OSCs, V_{oc} varies linearly with the HOMO level of the donor and the LUMO level of the acceptor. The I_{sc} is given by the density of absorbed photons per unit volume (*n*), the elementary charge (*e*), the device-dependent mobility (μ), and the electric field (*E*): $I_{sc} = ne\mu E$. Tests are carried out either under vacuum, in Ar or N_2, under AM 1.5G of white light irradiation by a Xe-lamp solar simulator.

The current–voltage (*I–V*) dependence of PV may be measured using any programmable instrument, capable of measuring *I* and *V* simultaneously such as, for example, a Keithley Source-Meter with the ability to measure voltage: 10 μV to 110 V and current: 10 fA to 100 mA.[134–136] The incident photon-to-current conversion efficiency can be measured with Stanford research systems (SRS) lock-in amplifier coupled with monochromator under dark and white light source, in which light intensity was calibrated at each wavelength, using a standard Si PV cell. Other light sources have also been used, such as tungsten-halogen with UV and IR filtered out or a quartz-halogen.[137,138] The measured I_{sc} values of PSCs need to be corrected to the values corresponding to the AM 1.5G reference. This is done by adjusting the solar simulator in such a way that: $I_{sc} = I_{sc,ref}/M$. Next, PCE, *FF* and other power conversion parameters may be calculated from Equation [3.3].[133,139,140]

Quantum efficiency spectra (EQE vs. λ) are measured using a lock-in amplifier for I_{sc} from short-circuited devices illuminated by chopped light from a tungsten-halogen lamp filtered by a monochromator. The incident power is determined using a calibrated photodiode, with typical power densities of $L = 0.1–1$ mWcm^{-2} and quantum efficiency $\mathrm{EQE}(\lambda) = 1240 I_{sc} / \lambda L$. The internal quantum efficiency (IQE) was determined from the amount of light absorbed by the blend via the ultraviolet–visible (UV-vis) measurements of the film on fused silica.[138]

It has been noted that the V_{oc} of PV containing thiophene-based polymers linearly increases with the onset of oxidation potential (OOP), proportional to the HOMO position of the donor polymer.[141,138] The OOP was determined electrochemically by cyclic voltammetry (CV). The experiments were carried out on thin films at room temperature (RT) in a glove box under dry Ar using a computer-controlled potentiostat with Pt-foils electrodes and Ag/AgCl coated, calibrated reference electrode (RE). The OOP was taken as the position where the oxidation current started deviating from the baseline. The accuracy of the measurement was confirmed using the electrochemical voltage spectroscopy (EVS).

The full range of microscopes has been used for characterization of PV morphologies. **AFM** is used for studying the morphology of polymer blends, surface of films cast from solutions with various solvents, as well as the blend structures at cross section.[98,142] After determining with AFM thickness of the film active layer and the surface morphology, the absorption spectra was measured using a UV-vis spectrophotometer. The tests were done under N_2.[138,143–145]

TEM has been used extensively for studying morphology and its evolution in polymer blends, including those employed in PV devices. Variation of polymer concentration, type of solvent, annealing, aging, etc., all affect the morphology and hence performance. TEM has an additional advantage as it allows performing selected-area electron diffraction (**SAED**). As an example, MDMO-PPV films prepared with 20–90 wt% PCBM were observed in TEM bright-field mode. SAED analysis showed that observed clusters consisted of PCBM nano-crystals with random crystallographic orientations. The TEM observations and acquisition of SAED patterns were conducted in a TEM operated at 80 kV. Annealing MDMO-PPV above its T_g accelerated the phase segregation and in the formation of large PCBM single crystals, thus indicating that the thermal stability may be the limiting factor of these SCs.[146]

SEM scans the specimen surface with a high-energy electron beam. The microstructures of the deposited PAB films are studied in SEM as film surface and microtomed cross sections. For enhancing the contrast between PAB phases, partial dissolution (by solvent vapor) or erosion (by plasma treatment) may be needed. Furthermore, for preventing accumulation of the static electricity on the specimen surface, a thin layer of Au/Pt may be deposited. The specimens are placed on a platinized silicon wafer holder with a conductive adhesive copper tape. The modern SEM has high resolution, revealing details less than 1 nm in size. With a large depth of field it gives a characteristic 3D appearance of the surface structure. [147]

UV-vis refers to absorption spectroscopy in the ultraviolet-visible spectral region. The absorption in the visible range directly affects the perceived color of the chemicals involved. The UV-vis spectra are measured using spin casted films, or dilute polymer solutions. Films are more representative for the active layer behavior in SCs, while solutions are more reliable when comparing the different polymers or blends. For example, the film deposited on ITO substrate may be dissolved in a suitable solvent (e.g., in chloroform at concentration of 25 μg/mL, in 10 mm quartz cell) and either directly used in a spectrometer, or spin casted on glass slides, vacuum dried and measured at the wavelength 280 to 900 nm at a rate of 300 nm/min.[135]

FT-IR with the attenuated total reflectance (ATR) is an analytical technique used for identifying mainly organic materials. ATR projects a wave into a specimen through a Ge, Si, ZnSe or diamond crystal. Thus the sample may be analyzed directly, without dissolving, or grounding with KBr.[148] The FT-IR/ATR system measures the absorption of various infrared light wavelengths by the material, identifying specific molecular dynamic elements and structures. The method is used primarily by synthetic chemists for identification of reaction products, although ^{1}H and ^{13}C NMR is often used as the principal method (see below).

NMR ^{1}H and ^{13}C spectra were measured for identification of the structure of a new *bi*-indene C_{70}-derivative (BC70MA; an acceptor) for PSCs with poly(3-hexylthiophene) (P3HT; a donor). Chemical shifts were reported in ppm relative to the singlet of CDCl3 at 7.26 and 77 ppm for ^{1}H NMR and ^{13}C NMR, respectively.[149,150] Woody *et al.* used FT-IR and ^{1}H NMR for determining the end groups of the donor-acceptor-donor (DAD) *tri*-block copolymers.[151]

XPS (X-ray photoelectron spectroscopy or ESCA = electron spectroscopy for chemical analysis) has been frequently used for the determination of local surface composition of a PV blend to the depth of <10 nm.[104,152]

X-ray diffraction **(XRD)** has been used for investigation of crystallinity in materials. The X-ray source may be any that generates X-rays, for example, synchrotron, cyclotron, X-ray tube with a monochromatic Cu-Kα_1 radiation, diffractometer and a programmable scanning capability with speed of around 1°/min. These instruments may be used for determining interlayer spacing within a crystalline material (polymers, carbon nanotubes, or minerals) and phase content of powders and coatings.[153,154]

SAXS (small-angle X-ray scattering) may be performed at a synchrotron facility. Powder specimens are loaded into glass capillaries and inserted into aluminum cassettes which had a 1.5 mm hole with ±15° angular range. The cassettes are inserted into a thermostatic stage mounted in a vacuum chamber and centered on the beam-line with the incident beam energy of 8 or 16 keV (λ = 1.55 or 0.775 Å, respectively). Two-dimensional SAXS images are recorded around 1.2 m from the sample at $T \leq 250 \pm 0.1$°C.[150]

TGA is used for determining volatiles content and/or thermal stability of materials, by monitoring the weight loss on heating (see Section 3.4.1).

SEC (previously gel permeation chromatography, **GPC**). The molecular weight and its polydispersity in polymers or copolymers are determined in SEC with Styragel or PSS SDV chromatographic columns, a gradient pump, automated sample injector and suitable detectors (e.g., UV/vis set to 250 nm, multi-angle laser light scattering, MALLS, at λ = 658 nm, refractive index, solution viscosity, and others). Traditionally the instrument is calibrated with anionic, narrow MWD PS standards.[135] For example, SEC and ^{1}H NMR spectroscopy were used for identifying the *tri*-block DAD copolymer structure.[151] Absolute molecular weights were determined by light scattering with dn/dc values calculated using the RI detector and assuming 100% mass recovery of the injected sample.[150]

3.7 Aerospace applications

Several books and reviews discussing the high temperature materials used in construction of aerospace vehicles are available.[155–157] The Society of Aerospace and Automotive Engineers (SAE) develops standards for materials, parts and construction of powered vehicles. Currently there are 434 active standards.[158] Table 3.3 provides an extracted list of examined polymers, their blends and composites. Evidently, these commercial materials are to be used not only in aerospace applications. However, many of the standards deal with high performance polymeric systems, for example, fluorosilicone thermoset elastomer with a range of operating temperatures from −85°C to +230°C, or high temperature engineering and specialty resins, blends and composites with glass or carbon fibers, etc.

3.7.1 HTPBs in the *Solar Impulse* aircraft

In 2003 Piccard and Borschberg from the École Polytechnique Fédérale de Lausanne (EPFL) initiated construction of the *Solar Impulse* – a fixed-wing aircraft that will circle the world on solar power.[159,160] The prototype, *Solar Impulse HB-SIA*, successfully passed all the tests. The advanced polymers used in its construction are listed in Table 3.4.[161]

Table 3.3 Summary of SAE polymers and blends standards (HTPs are in bold)[163]

Polymer	Abbreviation	Contains	Comment
Acrylonitrile butadiene rubber	NBR	Nylon cloth	Sheets, rings
Aramid fiber	PARA (LCP)	Polyester coated	Pultruded, rowing
Butyl rubber	IIR		Sheet, strip, tubing, extrusions, moldings
Cellulose acetate	CA		Moldings and extrusions
Cellulose acetate butyrate	CAB		Moldings and extrusions
Chloroprene rubber	Chloroprene (CR)	Nylon cloth	Gaskets, seals, diaphragms, strips
Epoxy cresol Novolac	ECN		For high temperature composite
Epoxy resin	EP	S-2 glass roving	Heat-curable strands
Epoxy resin	EP	Cycloaliphatic	For high temperature composite
Ethylene propylene rubber	EPR or EPDM		For seals, gaskets, grommets
Fluorocarbon rubber	FKM		O-rings, compression seals
Fluorosilicone rubber	FVMQ		Sheet, strip, tubing, extrusions, moldings
Melamine-formaldehyde	MF	Mineral filled	Moldings
Methylmetha-crylate	PMMA		Sheets, rods and tubes
Novolac epoxy resin	EPN	Short glass fiber	Moldings
Nylon	PA	PU coated	Braid, cloth, moldings, extrusion
Nylon/polyester	PA/PEST		Aircraft webbing
Perfluoroether elastomer	**PFPE**		Molded sealing rings
Phenolic resin	PH		Corrosion-preventive varnish
Polyamide-imide	PAI	PTFE, Cresol-formaldehyde (CF), GF,	Bar, rod, and shapes
Polybenzimidazole	PBI	PA, GF, PTFE	Impregnated, sintered
Polycarbonate	PC		Sheet
Polychloroprene	CR		Tubing, electrical insulation
Polychlorotrifluoro-ethylene	PCTFE		Film or tape, moldings
Polyester	PEST		Braid or film

Table 3.3 Continued

Polymer	Abbreviation	Contains	Comment
Polyether urethane	EU	Resin and hardener	Elastomeric sheets or foams
Polyethylene	PE		Rods and tubes
Polyethylene terephthalate	PET		Tubing, or insulation
Polyimide	PI	Glass or quartz cloth	Blocks, slices, shapes
Polyolefin	PO	Crosslinked, pigmented	Heat shrinkable, tubings
Polystryene	PS		Moldings and extrusions
Polysulfide (182 °C)	PTR		Adhesion promoter, sealing, lubricant
Polysulfone	PSF	CF, GF	Tape and sheet
Polytretrafluoro-ethylene (400°C)	PTFE	GF, CF	Extruded rods, tubes, profiles
Polytretrafluoro-ethylene/ fluorosilicone	PTFE/FVMQ		Molded seals
Polythioether rubber (182 °C)	PESK		Sealing compound
Polyurethane	PU		Reticulated foam
Polyvinyl chloride	PVC		Tubing, cord, tape, and shapes
Polyvinylidene fluoride	PVDF		Heat-shrinkable tubing
Silicone rubber	PVMQ		Sheet, strip, tubing, moldings, sponge
Silicone rubber	VMQ, curable	GF cloth or rowing	Gaskets or seals

Table 3.4 Material used in construction of *Solar Impulse HB-SIA*

Chemical name	Trade name
Aromatic polyamide, PARA	Ixef
Carbon fibers, CF	Low density
Perfluoropolyethers, PFPE	Fomblin
Poly(ethylene-co-chloro tri-fluoro ethylene), ECTFE	Halar
Polyamide-imide, PAI	Torlon
Polycarbonate, PC	Macrolon
Polyetheretherketone, PEEK	Keta Spire
Polyphenylene sulfone, PPSU	Radel
Polyphthalamide, PPA	Amodel
Polyvinylidene fluoride, PVDF	Solef
Self-reinforced polyphenylene ether, SRP	Primo-Spire

3.7.2 HTPBs in the IKAROS (Interplanetary Kite-craft Accelerated by Radiation of the Sun) spacecraft

The Japan Aerospace Agency (JAXA) experimental solar-sail spacecraft, IKAROS, was launched on 21 May 2010. Seven months later it passed Venus at a distance of about 80800 km, successfully completing the planned mission. As a reward, JAXA extended the mission to March 2012.[162] The spacecraft is in the form of a 14×14 m square sail, made of a 7.5 μm thick sheet of PI. Embedded in the sail, an array of amorphous silicon SCs provides the needed electrical power, while a liquid crystal device (LCD) controls the attitude by sunlight absorption/reflection.[163,164] One of the reasons for the mission's extraordinary success was the performance of the materials, especially the new PI ultra-thin film. Development of PI for the gossamer spacecrafts was described by Iwata *et al.* in 2009.[165] For the standard PI the high radiation resistance was in conflict with the thermoplasticity. In the new macromolecules, the authors incorporated non-symmetrical di-anhydrides and highly flexible aromatic ether units, which assured that the new PI are highly stretchable and resistant to radiation. For testing the material designed for deep space, the radiation durability was tested with proton beams under high vacuum.

3.7.3 HTPBs in the NASA polymeric spherical electrostatic shield

For several decades NASA evaluated diverse electrostatic shielding systems for spacecraft radiation protection.[166] The present passive shielding uses PE and is being employed on the International Space Station (ISS). The active shielding would generate the opposite charges, for example, positive to repel protons and negative to affect the low energy electrons. Thus the concept led to the design of three charged spheres surrounding the spacecraft. To avoid the huge penalty of the extra mass, the final step collapsed the spherical shields into charged spacecraft and six positively charged spheres (kept at +300 MV) and six negatively charged spheres (–300 MV) strategically placed around the spacecraft, in a pattern resembling a classic quadrupole (see Fig. 3.10). The keys to project success are the materials, configuration, power supply, and effectiveness compared to the passive shields. The polymer must be resistant to atomic oxygen (AO) and radiation (UV, VUV, γ, cosmic, galactic), as well as having suitable thermal, dielectric and mechanical properties. The polymers considered for the shielding spheres are listed in Table 3.5. The proposed conductive spheres would have four layers: TiO_2-filled PI, liquid crystal polymer (LCP) fabric, PI and a thin layer of Au on the

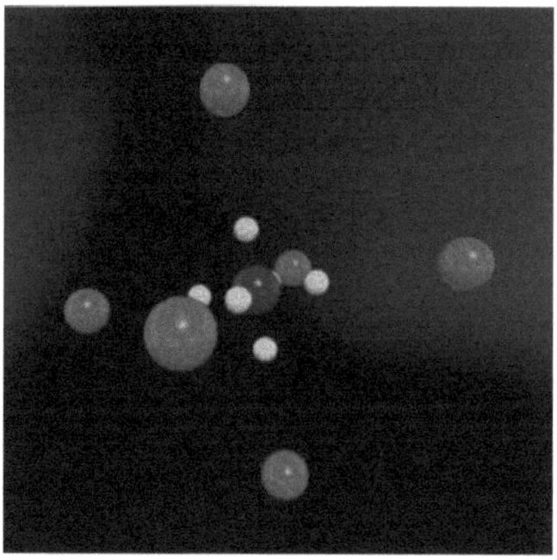

3.10 Positively and negatively charged spheres

Table 3.5 Room temperature DC dielectric and ultimate tensile strengths

Material		Source	Dielectric strength (kV/mm)		Tensile strength (MPa)
1	Kapton® HN	PI	DuPont	303	231
2	Kapton® FN1	PI	DuPont	272	207
3	Upilex®	PI film	UBE	272	362
4	Apical®	PI film	Kaneka	320	303
5	LaRC™ CP1	PI	SRS Technologies	ND	112
6	LaRC™ CP2	PI	SRS Technologies	ND	131
7	Ultem™	PEI	GE	33	127
8	TOR-LM	PAEBI*	Triton Systems	127	104
9	TOR-NC	PAEBI-F†	Triton Systems	190	114
10	Mylar®	PET	DuPont Teijin	433	216 (Molecular dynamics calculation (MD))
11	Teflon®	PTFE	DuPont	18 9	31
12	FEP Teflon®	FEP	DuPont	53	23

* TOR = Triton oxygen resistant, TOR-LM = low modulus co-poly(arylene ether benzimidazole); †TOR-NC is a fluorinated, triphenyl phosphine oxide based PBI.

interior. Figure 3.11 shows the structure of TOR-NC, a fluorinated, triphenyl phosphine oxide based PBI.

Trans-polyoctenamer rubber (TOR)-LM, TOR-NC, Upilex, the PI used in IKAROS, and several other polymers have been specifically

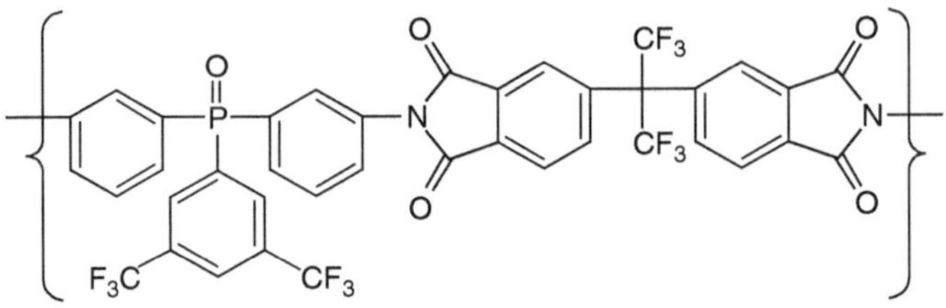

3.11 Chemical structure of TOR-NC.

designed for space.[167–171] Similarly, blends and filled thermoplastics or thermosets have been designed for such use, for example, blends of poly(vinylidene fluoride)/polyaniline (PVDF/PANI),[172] or polyether ether ketone/polysulfone (PEEK/PSU),[173] or polyimide/epoxy (PI/EP),[174] or self-lubricating MoS_2-filled PEEK.[175] In many cases the name hides the multi-ingredient nature of the polymer known by its trade name. In short, for most aerospace applications a combination of materials from Tables 3.3 to 3.5 is used – the choice depending on the function, expected lifetime and cost.

3.8 Characterization of HTPBs for aerospace applications

Aerospace materials are being tested for diverse purposes, for example, in research and development, for quality control, failure analysis, material certification, system design. The tests include analytical (chemical) and physical (mechanical, electrical, thermal, optical, and rheological) measurements as well as evaluation of flammability, physical and chemical (weathering).[176] Furthermore, these materials require tests that offer an insight into their behavior in the hostile environment of a low Earth orbit (LEO) as well as in deep space, namely, vacuum, radiation, large temperature differences and a combination of these factors. The high power lasers have found numerous uses for testing the spacecraft materials.[177]

LEO is defined as being at 160–2000 km above the Earth's surface (thus where the ISS and a majority of artificial satellites orbit the Earth). Objects in LEO are exposed to vacuum, UV and low density of atmospheric gases that on the one hand engender drag and on the other affect the stability of the polymeric components; for example, AO transforms PI surface into micrometer-sized roughness, known as a *carpet shape*.[178] Recently, 39 polymers were exposed to AO erosion for 3.95 years on an ISS surface (some samples for 5.8 years). The AO erosion was calculated from the weight loss and density, following the ASTM E 2089–00. Additional erosion tests performed on the ground are also described. As shown in Table 3.6, the

Table 3.6 Atomic oxygen erosion yield[178]

Material (abbreviation)	Erosion yield, 10–24 (mL/atom)
Crystalline, white polyvinyl fluoride (White PVF)	0.101
Polytetrafluoroethylene (PTFE Teflon)	0.142
Perfluoroalkoxy copolymer resin (PFA)	0.173
Amorphous fluoro polymer (AF)	0.198
Fluorinated ethylene propylene (FEP Teflon)	0.200
Pyrolytic graphite (PG)	0.415
Poly-(*p*-phenylene terephthalamide) (Kevlar, PPDT)	0.628
Chlorotrifluoroethylene (Kel-f, CTFE)	0.831
Polybutylene terephthalate (PBT)	0.911
Polyimide (biphenyl tetracarboxylic acid dianhydride, BPDA) (Uplex S)	0.922
Tetrafluorethylene-ethylene copolymer (Tefzel, ETFE)	0.961
Acrylonitrile butadiene styrene (ABS)	1.09
Polyvinylidene fluoride (Kynar, PVDF)	1.29
Poly(p-phenylene-2,6-benzobisoxazole) (Zylon, polybutyleneoxide (PBO))	1.36
Polyacrylonitrile (PAN)	1.41
Polyphenylene isophthalate (Nomex, PPPA)	1.41
Polyurethane (PU)	1.56
Ethylene-chlorotrifluoroethylene (Halar, ECTFE)	1.79
Polyamide 66 or nylon 66 (PA 66)	1.80
Polyamide (PI)	1.91
Polyethylene oxide (PEO)	1.93
Polybenzimidazole (PBI)	2.21
Polypropylene (PP)	2.68
Polyimide (PMDA) (Kapton HN, PI)	2.81
Polysulfone (PSU)	2.94
Polyetheretherketone (PEEK)	2.99
Polyimide (PMDA) (Kapton H, PI)	3.00
Polyethylene terephthalate (Mylar, PET)	3.01
High temperature polymide resin (PMR 15, PI)	3.02
Polyvinyl fluoride (Clear Tedlar, PVF)	3.19
Polyetherimide (PEI)	3.31
Polyamide 6 (PA 6)	3.51
Polyethylene (PE)	3.74
Polystyrene (PS)	3.74
Epoxy (Hysol EA 956, EP)	4.21
Polycarbonate (PC)	4.29
Cellulose acetate (CA)	5.05
Polymethylmethacrylate (PMMA)	5.60
Allyl *di*-glycol carbonate (CR-39, ADC)	6.80
Polyoxymethylene; acetal (Delrin, POM)	9.14

fluorinated polymers resisted the AO erosion best, while the polymers with a large quantity of oxygen the worst; polyimides (PI) such as Kapton® H performed rather poorly.

Miyazaki *et al.* examined a commercial (from Nippon Steel Chem.) polysiloxane-block-polyimide for its resistance to on-the-ground-performed AO erosion, based on the laser detonation system, which simulated a three month exposure on orbit at an altitude of 400 km. The erosion was compared to that of Kapton® H. The mass change measurement, cross-sectional TEM observation, and XPS analysis were performed. Results show that exposed to AO the block copolymer forms on its surface a self-organized layer of SiO_2 with a self-healing capability. As a consequence, the mass loss of the block copolymer was < 1/100 of that measured for Kapton® H, i.e., performing at least three times better than the best polymer of Table 3.6, the TiO_2-filled polyvinyl fluoride (PVF). It is noteworthy that the PVF with inorganic pigment had the AO erosion yield 32 times smaller than its neat, transparent PVF variety; after testing the 'white PVF' white powder was visible on the specimen surface.[179] These two observations resemble recent communication by Fuse *et al.*, in which the authors reported that incorporation of ≤2 wt% organoclay reduced the partial discharge erosion (PDE) depth of PP by about 50%.[180] Apparently, the initial vaporization of PP increased the clay concentration on the surface to a level that prevented further erosion. Interestingly, the initial rate of erosion linearly decreased with clay concentration (reduction by a factor of about nine for 6 wt% clay), but the final effect was virtually independent of the clay content.[181] A similar mechanism was also postulated for the flammability reduction of polymers after incorporation of clay.[182]

The AO erosion was found to be accelerated by radiation and vacuum ultraviolet (VUV) in LEO.[183] The ground-based experimental results were obtained using a hyperthermal AO beam generated by laser detonation. The combined effects of AO and VUV are important for fundamental and engineering points of view. The VUV-sensitive PMMA and fluorinated EPR (FEP) showed no synergistic effects.[184] In this case, the erosion effect was the sum of the AO and VUV effects. However, a year later Yuan *et al.* reported negative synergism of PMMA exposed to AO and VUV.[185] Miyazaki and his colleagues[165,171,186] also studied the AO and VUV erosion in LEO. The polysiloxane-*b*-polyimide film was exposed separately to an AO beam and a VUV beam at the JAXA Space Center in Tsukuba. Owing to formation of a protective SiO_2 layer, AO erosion was small. The VUV irradiation caused a worrisome specimen discoloration, but insignificant mass loss.

The authors provided detailed information on the test facility and method of testing at the JAXA test facility.

- The AO irradiation was performed using the laser detonation facility with fluence equivalent to three months exposure on 400 km altitude simulating ISS conditions. As an internal reference Kapton H was applied.

- The UV irradiation tests were performed using a 5 kW Xe short arc lamp with a dichroic mirror with intensity 10×solar constant. The air pressure was reduced to $P \sim 1.3\times10^{-3}$ Pa. The tests were carried out at wavelengths 0.2–0.4 μm, room temperature (RT) or $T \sim 100°C$, for 150 equivalent sun days (ESD). There was insignificant reduction of mechanical performance of irradiated PI evaluated near T_g. Alternatively, a deuterium (D_2) lamp with MgF_2 window may be used with wavelengths 0.112–0.400 μm. The specimen was mounted 30 cm from the sample and 15° from the surface normal, giving the output power of 350 μWcm^{-2}. A spectrometer was coupled to the vacuum chamber to perform *in situ* the infrared reflection-absorption spectroscopy (IRRAS). The *p*-polarized IR beam was directed onto the liquid-nitrogen-cooled mercury cadmium telluride (MCT) detector.[185]
- The surface of the irradiated films was observed in an AFM with silicon nitride tips at RT. The AFM imaged 10×10 μm areas from several spots of each specimen. The images were analyzed plotting the pixel heights vs. the lateral dimensions. The surface roughness was computed as the number and weight averages.
- The proton irradiation tests were carried out in a Tandem accelerator under vacuum ($P < 2.6\times10^{-4}$ Pa) with the energy of 2.6 MeV; an irradiation dose of 63 MGy reduced the elongation at break of PI by around 50 % of initial value.
- A Cockcroft-Walton accelerator was used for the electron irradiation under vacuum ($P < 7.9\times10^{-6}$ Pa) with accelerating potential of 200–500 kV. After the dose of 20 MGy, the decrease of elongation at break was comparable to that of proton irradiation.
- Spacecrafts in orbit are irradiated by electrons from space, which accumulate on the satellite surface. The differential charging of the external dielectrics may be as high as several kV, causing discharges and anomalies. The conductivity of space-used materials should be known for the modeling of satellite charging, but suitable methods have not been developed. A procedure was proposed for charging grounded dielectrics with one electron beam and measuring the charge decay with another.[187] Once the charge decay rate is determined, the material conductivity may be calculated. The procedure is versatile for simulating space conditions (dose rate, total dose, potential, material aging, etc.).
- Mechanical testing was performed on epoxy resin toughened by incorporation of up to 17 vol% of PEI.[188] The dynamic mechanical tests were carried out in DMTA at 10 Hz, in a single cantilever bending mode at $T = -80$ to 300°C. The flexural properties at 23°C were carried

out using the supporting span of 48 mm (ASTM D7907; specimens: 60×10×3 mm).

- Fracture mechanics measurements were made according to ASTM E3998 at 23°C in three-point bending at a crosshead speed of 1.3 mm/min, on sharply notched rectangular bars. Fractured surfaces were Au/Pl coated and examined in SEM.[188]
- Fracture toughness was determined according to ASTM E-399 using 1× 1× $\frac{1}{8}$ inch compact-tension epoxy specimens toughened with dispersed LCP whiskers.[189] The latter were obtained by blending the LCP [poly(ethylene terephthalate-*co*-*p*-oxybenzoate), PET/PHB60] with Ardel-D100 (polyester of bisphenol-A and a mixture of tere- and isophthalic acids), dissolving the blend in Epon 825, curing it and causing crystallization of LCP into high aspect ratio fibrillar crystals.
- ASTM D5868 – 01(2008) Lap Shear Adhesion (LSA) test for fiber reinforced plastic (FRP) bonding was used for evaluating the surface modification of PBI (service temperature: –260 to +400°C) by exposing it to high-energy radiation and low-pressure plasma. The treatment increased LSA 20-fold.[170] The surface energy of PBI was measured by the contact angle method.
- Small angle neutron scattering (SANS) studies in Grenoble used a cold source of neutrons of wavelength 1 nm at high flux.[168] The curing samples, housed in a hot stage holder, permit observation of the morphology evolution. Plots of scattering intensity vs. the scattering vector, Q, offer an insight into the sample's morphology. For example, a peak in the dependence indicates a regular structure, an interpenetrating polymer network (IPN), or a co-continuous morphology, such as formed during the spinodal decomposition (SD). The authors found significant enhancement of epoxy toughness when the disordered morphology of dispersed polyether sulfone (PES) changed into the SD-type.
- For the identification of space effects on polymers, numerous standard tests must be used, namely spectroscopy, thermal analysis, dielectric and conductivity measurements, mechanical and viscoelastic property determination, etc. In the late 1990s NASA decided to substitute polyurethane coatings (having carcinogenic) with another one.[189,190] The 18 alternative candidates were subjected to 34 standard test procedures (Table 3.7). In the following report the number of candidates fell to eight, from which the best with the lowest cost-to-performance ratio will have to be selected.[191] This procedure illustrates how involved the polymeric testing for space application is, even if the test procedures are quite standard.

Table 3.7 Recommended test procedures for replacement aliphatic PU coatings[192]

Screening tests for alternatives to aliphatic isocyanate urethane coatings	Common tests for alternatives to aliphatic isocyanate urethane coatings
Pot Life (Viscosity); ASTM D 1200	Removability; ASTM G 155
Ease of Application; SSPC-PA-2	Repairability; ASTM D 523, 2244 and 3359
Surface Appearance; ASTM D 523 and 2244	Abrasion Resistance; ASTM D 4060
Dry-To-Touch (Sanding); none	Gravelometer; ASTM D 3170
Accelerated Storage Stability; ASTM D 1849	Fungus Resistance; ASTM D 3359 and MIL-STD-810F
Cure Time (MEK Solvent Rub); ASTM D 4752	Accelerated Weathering; ASTM D 523, D 2244 and G 155
Solvent (Acetone) Rub; ASTM D 4752	Mandrel Bend Flexibility; ASTM D 522
Cleanability; MIL-PRF-83282D and 85285	18-Month Marine Environment; ASTM D 610, 714 and 523
X-Cut Adhesion by Wet Tape; ASTM D 3359 and FED-STD-141	Cyclic Corrosion Resistance; GM 4465 P and 9540 P
Tensile (Pull-Off) Adhesion; ASTM D 4541	Hypergol Compatibility; KSC MTB-175–88
Knife Test; FED-STD-141	LOX Compatibility; ASTM D 2512 and NASA-STD-6001

Notes: KSC, Kennedy Space Center; LOX, liquid oxygen tanks.

3.9 Summary

This chapter has described the application of HTPBs in fuel and solar cells as well as in aerospace. The three categories of HTP constitute but a small portion of the total polymer production, namely commodity, engineering, specialty and ultra categories, 95.29, 4.5, 0.2 and 0.01 wt%, respectively. By adjusting the molecular parameters and incorporating suitable additives, the commercial polymers may be formulated for a variety of other applications.

A very special category of characterization methods involves the use of HTPBs in highly demanding engineering applications such as, for example, structured membranes for gas or liquid separation at high temperature, *T*, and pressure, *P*, the fuel cell membranes and active layers in solar cells. Even more demanding are the aerospace applications, where the polymers may be exposed to wide *T* and *P* variations, hard radiation and erosion by atomic oxygen. As discussed, evaluation of the material performance for these applications requires elaborate specific test instrumentation and methods, often developed and maintained by a national organization.

The use of polymers and their blends in such areas as membrane applications is in a relatively early stage, it is expensive, and lacking efficiency and durability. In particular, for fuel cells there are intense efforts for replacing expensive Pt catalysts and polymeric per-fluoro sulfonic acid (Nafion®). Highly promising developments of these materials have been announced.[193–195] Other work has focused on expanded use of PAB in electronic (e.g., as organic transistors) and optoelectronic (photovoltaics, LED) applications, as conductive material (plastic wires), bio-materials and others.[196,197] Nano-particle technology is expanding to fuel and solar cell membranes. As in other fields, replacement of nano-size spheres by rods and plates offers significant improvement at low concentration.[197,198] In addition, nanoparticles influence crystallization of the semicrystalline polymer in its vicinity.[199] Layered clay nanoparticles, layered graphene, carbon nanotubes, metallic or metal oxide nanoparticles are used industrially. Hopefully, collaboration between the ceramic and polymer industries will yield a new family of synthetic, high temperature resistant nanoparticles with affinity to specific polymers that will eliminate the need for intercalation or compatibilization.[153,200,201]

3.10 References

1. Kemmish, D. J. High Performance Engineering Plastics, RAPRA Review Reports, Report 86, iSmithers Publishing, Vol. 8, No. 2 (1995).
2. Shannon, M. A.; Bohn, P. W.; Elimelech, M.; Georgiadis, J. G.; Mariñas, B. J. and Mayes, A. M., Science and technology for water purification in the coming decades, *Nature*, **452**(20), 301–310 (2008).
3. Norman, N. U.; Fane, A. G.; Winston Ho, W. S. and Matsuura, T., Eds., *Advanced Membrane Technology and Applications*, John Wiley & Sons, Inc., Hoboken, NJ, 994 pp. (2008); ISBN: 978–0-471–73167–2.
4. Dow, Water and Process Solutions FILMTEC™ Reverse Osmosis Membranes, Technical Manual, 181 pp., Form No. 609–00071–1009.
5. Macedonio, F. and Drioli, E., Membrane engineering progresses in desalination and water reuse, *Membrane Water Treatment*, **1**(1), 75–81 (2010).
6. Drioli, E. and Macedonio, F., Membrane Research, Membrane Production and Membrane Application in China, Report prepared in the framework of the Section on Membrane Engineering of the European Federation of Chemical Engineering (EFCE), Institute on Membrane Technology ITM-CNR, Università della Calabria, Italy (2008); www.itm.cnr.it/data/section/CHINA_Report.pdf - 2011–05–04
7. Park, Y. G., Effect of hydraulic pressure and electric field on electro-microfiltration in the wine-brewery industry, *Desalination*, **191**(1–3), 411–416 (2006).
8. Xiao, Y.; Low, B. T.; Hosseini, S. S.; Chung, T. S. and Paul, D. R., The strategies of molecular architecture and modification of polyimide-based membranes for CO2 removal from natural gas-A review, *Prog. Polym. Sci.*, **34**, 561–580 (2009); DOI:10.1016/j.progpolymsci. 2008.12.2004.
9. Abedini, R. and Nezhadmoghadam, A., Application of membrane in gas separation processes: its suitability and mechanisms, *Petroleum & Coal*, **52**(2) 69–80 (2010); Available online at www.vurup.sk/pc.

10. Drioli, E. and Barbieri, G. (Eds), *Membrane Engineering for the Treatment of Gases: Gas-separation Problems with Membranes*, Royal Society of Chemistry, 2011, ISBN: 978-1-84973-244-4.
11. Pientka, Z. and Schauer, J., Hydrogen economy and polymer membranes, *Macromol. Symp.* (Special Issue: New Frontiers in Macromolecular Science), **295**(1), 23–29 (2010).
12. Koltuniewicz, A. B. and Drioli, E., *Membranes in Clean Technologies, Theory and Practice*, Wiley-VCH, Weinheim, Bergstr: May 2008, ISBN-10: 3–527–32007–5.
13. Li, Q.; Jensen, J. O.; Savinell, R. F. and Bjerrum, N. J., High temperature proton exchange membranes based on polybenzimidazoles for fuel cells, *Prog. Polym. Sci.*, **34**, 449–477 (2009); DOI:10.1016/j.progpolymsci.2008.12.003.
14. Asensio, J. A.; Sánchez, E. M. and Gómez-Romero, P., Proton-conducting membranes based on benzimidazole polymers for high-temperature PEM fuel cells. *A Chemical Quest, Chem. Soc. Rev.*, **39**, 3210–3239 (2010); DOI: 10.1039/b922650h
15. Veenstra, S. C.; Loos, J. and Kroon, J. M., Nanoscale structure of solar cells based on pure conjugated polymer blends, *Prog. Photovolt: Res. Appl.*, **15**, 727–740 (2007); DOI: 10.1002/pip.796.
16. Bilby, D.; Kim, B. G. and Kim, J.-S., Recent design strategies for polymer solar cell materials, *Pure Appl. Chem.*, **83**(1), 127–139 (2011); DOI:10.1351/PAC-CON-10–08–23.
17. Hosseini, S. S.; Teoh, M. M. and Chung, T. S., Hydrogen separation and purification in membranes of miscible polymer blends with interpenetration networks, *Polymer*, **49**, 1594–1603 (2008); DOI: 10.1016/j.polymer.2008.01.052.
18. Hosseini, S. S.; Peng, N. and Chung, T. S., Gas separation membranes developed through integration of polymer blending and dual-layer hollow fiber spinning process for hydrogen and natural gas enrichments, *J. Membr Sci.*, **349**, 156–166 (2010); DOI: 10.1016/j.memsci.2009.11.043.
19. Camacho-Zuniga, C.; Ruiz-Trevino, F. A.; Hernandez-Lopez, S.; Zolotukhin, M. G.; Maurer, F. and A. Gonzalez-Montie, A., Aromatic polysulfone copolymers for gas separation membrane applications, *J. Membr. Sci.*, **340**(1–2), 221–226 (2009).
20. Li, Y.; Cao, C.; Chung, T. S. and Pramoda, K. P., Fabrication of dual-layer polyethersulfone (PES) hollow fiber membranes with an ultrathin dense-selective layer for gas separation, *J. Membr. Sci.*, **245**, 53–60 (2004); DOI: 10.1016/j.memsci.2004.08.002.
21. Corbeau, A. S., Piles à Combustible – Hydrogène, 2000–2008, http://www.annso.freesurf.fr/type.html.
22. Waidhas, M.; Drenckhahn, W.; Preidel, W. and Landes, H., Direct-fuelled fuel cells, *J. Power Sources*, **61**(1–2), 91–97 (1996).
23. Steinberg, M., Application of the natural gas direct carbon fuel cell (NGDCFC) to a gas filling station for hydrogen and electricity supply, *Energy Procedia*, **1**(1), 1427–1434 (2009).
24. Jones, D., Membrane development for medium and high temperature PEM-FC in Europe, Proceedings of the HTMWG-CARISMA session Brussels 10–11.10.2007; http://www1.eere.energy.gov/hydrogenandfuelcells/pdfs/htwmg_oct07_jones.pdf.
25. Pivovar, B., A discussion on improved HTMs, High Temperature Membrane Working Group (HTMWG) Meeting, Honolulu, HI, 16.10.2008; http://www1.eere.energy.gov/hydrogenandfuelcells/pdfs/htmwg_oct08_pivovar.pdf

26. http://www.hydrogen.energy.gov/pdfs/9012_fuel_cell_system_cost.pdf.
27. Greszler, T., Membrane performance and durability overview for automotive fuel cell applications, *Presentation at 2006 High Temperature Membrane Working Group Meeting*, San Francisco, CA., 14 September 2006.
28. Leikin, A. Yu.; Bulycheva, E. G.; Rusanov, A. L. and Likhachev, D. Yu., High-temperature proton-exchange membranes based on polymer–acid complexes, *Vysokomol. Soed., B*, **48**(6), 1031–1040 (2006).
29. Rusanov, A. L.; Likhatchev, D. Yu. and Müllen, K., Proton-conducting electrolyte membranes based on aromatic condensation polymers, *Russ. Chem. Rev.*, **71**(9), 761–774 (2002).
30. Gubler, L. and Scherer, G. G., Strategy for aging tests of fuel cell membranes, HTMWG – CARISMA, High Temperature Membrane Working Group Meeting, Washington D. C., 10 October 2007.
31. Chung, T. S.; Chen, P. N. Sr., Polybenzimidazole (PBI) and polyarylate blends, *J. Appl. Polym. Sci.*, **40**, 1209–1222 (1990).
32. Kallitsis, J. K.; Gourdoupi, N., Proton conducting membranes based on polymer blends for use in high temperature PEM fuel cells, *J. New Mater. Electrochem. Syst.*, **6**(4), 217–222 (2003).
33. Nam, K.; Xu, H.; Cao, S.; Olmeijer, D.; Servaites, J.; Wang, Y., Acid-base proton conducting polymer blend membrane, Int. Pat. Appl. WO 2003/062493, 23.01.2003, to Polyfuel, Inc.
34. Allcock, H. R.; Hofmann, M. A.; Ambler, C. M.; Maher, A. E.; Wood, R. M.; Welna, D. T., Synthesis of polyphosphazenes with sulfonimide side groups, US Pat. Appl., 20040225153, 11.11.2004, to The Penn State Research Foundation, University Park, PA.
35. Li, Q.; He, R.; Jensen, J. O.; Bjerrum, N. J., PBI-based polymer membranes for high temperature fuel cells – Preparation, characterization and fuel cell demonstration, *Fuel Cells*, **4**(3), 147–159 (2004); DOI: 10.1002/fuce.200400020.
36. Hu, K.; Xu, T. W.; Yang, W. H.; Fu, Y., Preparation of novel heterogeneous cation-permeable membranes from blends of sulfonated poly(phenylene sulfide) and poly(ether sulfone), *J. Appl. Polym. Sci.*, **91**(1), 167–174 (2004).
37. Swier, S.; Ramani, V.; Fenton, J.M.; Kunz, H.R.; Shaw, M.T.; Weiss, R. A., Polymer blends based on sulfonated poly(ether ketone ketone) and poly(ether sulfone) as proton exchange membranes for fuel cells, *J. Membr. Sci.*, **256**, 122–133 (2005).
38. Swier, S.; Chun, Y. S.; Gasa, J.; Shaw, M.T. and Weiss, R. A., Sulfonated poly(ether ketone ketone) ionomers as proton exchange membranes, *Polym. Eng. Sci.*, **45**,1081–1091 (2005).
39. Gasa, J. V.; Weiss, R. A.; Shaw, M. T., Influence of blend miscibility on the proton conductivity and methanol permeability of polymer electrolyte blends, *J. Polym. Sci. Part B: Polym. Phys.*, **44**, 2253–2266 (2006); DOI: 10.1002/polb.20865.
40. Gasa, J. V.; Boob, S.; Weiss, R. A.; Shaw, M. T., Proton-exchange membranes composed of slightly sulfonated poly(ether ketone ketone) and highly sulfonated crosslinked polystyrene particles, *J. Membr. Sci.*, **269**, 177–186 (2006).
41. Lavorgna, M.; Mensitieri, G.; Scherillo, G.; Shaw, M. T.; Swier, S.; Weiss, R. A., Polymer blend for fuel cells based on SPEKK: effect of co-continuous morphology on water sorption and proton conductivity, *J. Polym. Sci. Part B: Polym. Phys.*, **45**, 395–404 (2007).
42. Okazaki, Y.; Nagaoka, S.; Kawakami, H., Proton-conductive membranes based on blends of polyimides, *J. Polym. Sci. Part B: Polym. Phys.* **45**, 1325–1332 (2007).

43. Gasa, J. V.; Weiss, R. A.; Shaw, M. T., Structured polymer electrolyte blends based on sulfonated polyetherketoneketone (SPEKK) and poly(ether imide) (PEI), *J. Membr. Sci.*, **320**, 215–223 (2008).
44. Kim, D. H. and Kim, S. C., Transport properties of polymer blend membranes of sulfonated and nonsulfonated polysulfones for direct methanol fuel cell application, *Macromol. Res.*, **16**(5), 457–466 (2008).
45. Lee, B. J.; Sounai, A., Polymer electrolyte membrane for fuel cell and membrane-electrode assembly and fuel cell including the same, US Pat. Appl., 20080305379, 11.12.2008, to Cheil Industries Inc.
46. Cho, C. G.; Kim, S. H.; Park, Y. C.; Kim, H.; Park, J.-W., Fuel cell membranes based on blends of PPO with poly(styrene-b-vinyl-benzyl-phosphonic acid) copolymers, *J. Membr. Sci.*, **308** (1–2), 96–106 (2008).
47. Fu, R.-Q.; Julius, D.; Hong, L.; Lee, J.-Y., PPO-based acid-base polymer blend membranes for direct methanol fuel cells, *J. Membr. Sci.*, **322**(2), 331–338 (2008).
48. Gourdoupi, N.; Daletou, M.; Neophytides, S.; Kallitsis, J., High temperature polymer electrolyte membranes and membrane electrode assemblies based on blends of aromatic polyethers, International Application WO 2008/032228 a2, 20.03.2008, to Advent Technologies.
49. Gourdoupi, N.; Andreopoulou, A. K.; Deimede, V.; Kallitsis, J.K., A novel proton conducting polyelectrolyte composed of an aromatic polyether containing main-chain pyridine units for fuel cells applications, *Chem. Mater.*, **15**(26), 5044–5050 (2003).
50. Kallitsis, J.K.; Geormezia, M.; Neophytides, S., Polymer electrolyte membranes for high-temperature fuel cells based on aromatic polyethers bearing pyridine units, *Polym. Int.*, **58**, 1226–1233 (2009).
51. Pefkianakis, E. K.; Morfopoulou, C.; Deimede, V.; Kallitsis, J. K., Blends of aromatic polyethers bearing polar pyridine units and their evaluation as high temperature polymer electrolytes, *Macromol. Sympos.*, **279**(1), 183–190 (2009).
52. Deng, S.; Mays, J. W.; Hassan, M. K.; Mauritz, K. A., Low cost high temperature fuel cell membranes based on poly(1,3-cyclohexadiene) homopolymers, polymer blends, and block copolymers, the 238th ACS National Meeting, ACS Division of Fuel Chemistry, Washington, DC, 16–20.08.2009.
53. Deng, S.; Hassan, M. K.; Mauritz, K. A.; Mays, J. W. Gido, S. P., Materials comprising polydienes and hydrophilic polymers and related methods, US Pat. Appl., 2010/022556, 28.10.2010, to University of Tennessee Research Foundation.
54. Daletou, M. K.; Kallitsis, J. K.; Voyiatzis, G. and Neophytides, S. G., The interaction of water vapors with H3PO4 imbibed electrolyte based on PBI/polysulfone copolymer blends, *J. Membr. Sci.*, **326**, 76–83 (2009).
55. Li, Q. F.; Rudbeck, H. C.; Chromik, A.; Jensen, J. O.; Pan, C.; Steenberg, T.; Calverley, M.; Bjerrum, N. J.; Kerres, J., Properties, degradation and high temperature fuel cell test of different types of PBI and PBI blend membranes, *J. Membr. Sci.*, **347**(1–2), 260–270 (2010).
56. Gourdoupi, N.; Kallitsis, J. K.; Neophytides, S., New proton conducting polymer blends and their fuel cell performance, *J. Power Sources*, **195**(1), 170–174 (2010).
57. Thomas, O. D.; Peckham, T. J.; Thanganathan, U.; Yang, Y.; Holdcroft, S., Sulfonated polybenzimidazoles: proton conduction and acid–base crosslinking, *J. Polym. Sci. Part A: Polym. Chem.*, **48**, 3640–3650 (2010); DOI: 10.1002/pola.24147

58. Wang, Y.-F.; Hsu, T.; Hay, A. S.; Li, K.; Patel, B., High temperature polymer blends of poly(aryl ether ketone phthalazinone), US Pat. Appl., 20110104417 A1, 05.05.2011, to Polymics, Ltd., State College, PA (USA).
59. Sen, U.; Acar, O.; Bozkurt, A. and Ata, A., Proton conducting polymer blends from poly(2,5-benzimidazole) and poly(2-acrylamido-2-methyl-1-propane-sulfonic acid), *J. Appl. Polym. Sci.*, **120**(2), 1193–1198 (2011); DOI: 10.1002/app.33026.
60. Acar, O.; Sen, U.; Bozkurt, A. and Ata, A., Blend membranes from poly(2,5-benzimidazole) and poly(styrene sulfonic acid) as proton-conducting polymer electrolytes for fuel cells, *J. Mater. Sci.,* **45**: 993–998 (2010); DOI 10.1007/s10853-009-4030-6.
61. Hu, J.; Conrad, O., Proton exchange membrane comprising polymer blends for use in high temperature proton exchange membrane fuel cells, Int. Pat. Appl. WO 2011/035795, 31.03.2011, to Ewe-Forschungszentrum Für Energietechnologie E. V.
62. Lovato, T.; McNabb, K.; Meyers, J.; Rupnik, G.; Ma, Q.; Zhao, M.; Pan, J.; Walker, J.; Smith, T.; Cebe, P., Polymer Blends for Proton Exchange Membranes, Bull. Am. Phys. Soc., APS Meeting, 21–25.03.2011; Dallas, Texas, 56(1), S1.00039.
63. Fan, Z., Hyperbranched aromatic polyesters and their application in blends of linear polyamides, Doctor rerum naturalium dissertation, Technischen Universität, Dresden (August, 2009).
64. Zhang, X., Preparation and characterization of proton exchange membranes for direct methanol fuel cells, PhD Thesis, Universitat Rovira i Virgili (EUI), Tarragona, Spain, 17 November 2005.
65. Kitayama, T. and Hatada, K., NMR Spectroscopy of Polymers, Springer Laboratory 2004, ISBN 978-3-540-40220-6.
66. Utracki, L. A., On the computation of viscosity measurements data of dilute solutions of high polymers, *J. Polym. Sci. Part A-1: Polym. Chem.*, **4**(3), 717–721 (1966).
67. Utracki, L. A. *Rheology of Polymer Blends, in Encyclopedia of Polymer Blends*, Chapter 2 in Volume 2, Isayev, A., Ed., Wiley-VCH Verlag, Weinheim, Germany (2011).
68. Utracki L. A., Polymer Alloys and Blends, Hanser Verlag, Munich, New York (1989); revised version, translated into Japanese by Prof. T. Nishi was published by Tokyo Kagaku Dozin Co. Ltd., Tokyo (1991).
69. Utracki L. A., Ed., *Polymer Blends Handbook*, 2 Volumes, Kluwer Academic Publications, Dordrecht (2002).
70. Olabisi, O.; Robeson, L. M. and Shaw, M. T. *Polymer-Polymer Miscibility*, Academic Press, New York (1979).
71. Ehrenstein, G.; Engel, L.; Klingele, H. and Schaper, H., *Scanning Electron Microscopy of Plastics Failure*, Hanser Verlag, Munich (2010); ISBN: 9781569905029.
72. Vielstich, W.; Yokokawa, H. and Gasteiger, H. A., Eds., *Handbook of Fuel Cells: Fundamentals, Technology and Applications*, John Wiley &Sons, Ltd., Chichester, UK (2009).
73. Teo, W. K.; Li, K. and Wang, D., Highly permeable polyethersulfone hollow fiber membranes for gas separation, US Pat., 6,017,474, 25.01.2000, to National University of Singapore.

74. Wang, D. L.; Teo, W. K. and Li, K., Preparation and characterization of high-flux polysulfone hollow fibre gas separation membranes, *J. Membr. Sci.*, **204** 247–256 (2002).
75. Robeson, L. M., *Polymer Blends; A Comprehensive Review*, Hanser Publishers, Munich, Germany, (2007); pg. 358.
76. Freeman, B. D., Basis of permeability/selectivity tradeoff relations in polymeric gas separation membranes, *Macromolecules*, **32**, 375–380 (1999).
77. Robeson, L. M., Polymer blends in membrane transport processes, *Ind. Eng. Chem. Res.*, **49**, 11859–11865 (2010); DOI: 10.1021/ie100153q.
78. Ketola, R. A.; Kiuru, J. T.; Tarkiainen, V.; Kotiaho, T. and Sysoev, A. A., Comparison of analytical performances of a micro-array quadrupole instrument and a conventional quadrupole mass spectrometer equipped with membrane inlets, *Rapid Comm. Mass Spectrom.*, **17**(7), 653–756 (2003); DOI: 10.1002/rcm.973.
79. Fang, J. and Shen, P.K., Quaternized poly(phthalazinone ether sulfone ketone) membrane for anion exchange membrane fuel cells, *J. Membr. Sci.*, **285**, 317–322 (2006).
80. Ren, X.; Wilson, M. S. and Gottesfeld, S., High performance direct methanol polymer electrolyte fuel cells, *J. Electrochem. Soc.*, **143**(1), L12–L15 (1996).
81. Ma, S.; Kuse, A.; Siroma, Z. and Yasuda, K., Measuring conductivity of proton conductive membranes in the direction of thickness, *Espec Technology Report*, **20**, 12–20 (2005); Ma, S.; Tanaka, H., Measuring conductivity of proton conductive membranes in the direction of thickness, Part II, *Espec Technology Report*, **23**, 1–15 (2007).
82. Jiang, F.; Kaltbeitzel, A.; Fassbender, B.; Brunklaus, G.; Pu, H.; Meyer, W. H.; Spiess, H. W. and Wegner, G., Effect of polymer composition and water content on proton conductivity in vinyl benzyl phosphonic acid – 4-vinyl pyridine copolymers, *Macromol. Chem. Phys.*, **209**, 2494–2503 (2008); DOI: 10.1002/macp.200800375.
83. Titvinidze, G.; Kaltbeitzel, A.; Manhart, A. and Meyer, W. H., Synthesis and characterisation of sulphonated poly(arylene sulphone) terpolymers with triphenylphosphine oxidemoieties for proton exchange membrane fuel cells, *Fuel Cells*, **10**(3), 390–400 (2010); DOI: 10.1002/fuce.200900151.
84. Jiang, F., Proton-conducting copolymers, blends and composites with phosphonic acid as protogenic group, Dissertation zur Erlangung des Grades Doktor der Naturwissenschaften, Johannes Gutenberg-Universität Mainz (2009).
85. Ngai, K. L., Dynamic and thermodynamic properties of glass-forming substances, *J. Non-Crystalline Solids*, **275**, 7–51 (2000).
86. Protsailo, L., Development of high temperature membranes and improved cathode catalysts, Report to UTC Fuel Cells, Project DOE # DE-FC04–02AL67608, 01.01.2002 –31.12.2005; http://www.osti.gov/bridge/servlets/purl/882933-MDt3XD/882933.pdf.
87. Kamanyi, A. E.; Grill, W.; Ngwa, W. and Luo W., Advances in phase-sensitive acoustic microscopy studies of polymer blend films: annealing effects and micro-elastic characterization of PS/PMMA blends, *J. Microsc.*, **238**(2),134–144 (2010).
88. Chapin, D. M.; Fuller, C. S. and Pearson, G. L., A new silicon p–n junction photo cell for converting solar radiation into electrical power, *J. Appl. Phys.*, **25**, 676–677 (1954).

89. Honsberg, C. and Barnett, A., (Electrical and Computer Engineering, University of Delaware) Achieving a solar cell of greater than 50%: Physics, technology, implementation and milestones, Distinguished Lecture at UC Santa Barbara, 1 April 2008.
90. NREL Solar Cell Sets World Efficiency Record at 40.8%; http://www.typepad.com/services/trackback/6a00d8341c4fbe53ef00e553ff10688834.
91. http://optics.org/news/1/5/5.
92. Ghosh, A. K. and Feng, T., Merocyanine organic solar cells, *J. Appl. Phys.*, **49**(12), 5982–5989 (1978).
93. Whitlock, J. B.; Panayotatos, P.; Sharma, G. P.; Cox, M. D.; Sauers, R. R. and Bird, G. R., Investigations of materials and device structures for organic semiconductor solar cells, *Opt. Eng.*, **32**(8), 1921–1934 (1993).
94. Ebrahim, S. M.; Soliman, M. M. and El-Latif, M. M. A., Blend of Nylon 6 and polyaniline doped with sulfanilic acid and its Schottky diode, *High Perform. Polym.*, **22**, 377–391 (2010); DOI: 10.1177/0954008309103796.
95. Green, M. A.; Emery, K.; Hishikawa, Y. and Warta, W., Solar cell efficiency tables (Version 33), *Prog. Photovolt: Res. Appl.*, **17**, 85–94 (2009); DOI: 10.1002/pip.880.
96. Gregg, B. A. and Hanna, M. C., Comparing organic to inorganic photovoltaic cells: theory, experiment, and simulation, *J. Appl. Phys.*, **93**(6), 3605–3614 (2003).
97. Lloyd, M. T.; Olson, D. C.; Garcia, A.; Kauvar, I.; Kopidakis, N.; Reese, M. O.; Berry, J. J. and Ginley, D. S., Understanding Degradation Pathways in Organic Photovoltaicsµ, Poster Presented at the 2011 Photovoltaic Module Reliability Workshop, 16–17 February 2011, Golden, CO. NREL (National Renewable Energy Laboratory); Report #: PO-5200–50739. See also: Lloyd, M. T.; Garcia, A.; Berry, J. J.; Reese, M. O.; Ginley, D. S.; Olson, D. C., Overcoming degradation in organic photovoltaicsµ in: *IEEE Proceedings 37th IEEE Photovoltaic Specialist Conference*, 19–24 June 2011, Washington State Convention Center, Seattle, WA (2011).
98. Arias, A. C.; MacKenzie, J. D.; Stevenson, R.; Halls, J. J. M.; Inbasekaran, M.; Woo, E. P.; Richards, D. and Friend, R. H., Photovoltaic performance and morphology of polyfluorene blends: a combined microscopic and photovoltaic investigation, *Macromolecules*, **34**, 6005–6013 (2001).
99. Tang, C. W., Two-layer organic photovoltaic cell, *Appl. Phys. Lett.*, **48**(2), 183–185 (1986).
100. Koetse, M. M.; Sweelssen, J.; Hoekerd, K. T.; Schoo, H. F. M.; Veenstra, S. C.; Kroon, J. M.; Yang, X. and Loos, J., Efficient polymer: polymer bulk heterojunction solar cells, *Appl. Phys. Lett.*, **88**, 083504 (2006); DOI:10.1063/1.2176863.
101. Venkataraman, D.; Yurt, S.; Venkatraman, B. H. and Gavvalapalli, N., Role of molecular architecture in organic photovoltaic cells, *J. Phys. Chem. Lett.*, **1**, 947–958 (2010); DOI: 10.1021/jz1000819.
102. Topham, P. D.; Parnell, A. J. and Hiorns, R. C., Block copolymer strategies for solar cell technology, *J. Polym. Sci., Part B: Polym. Phys.*, **49**, 1131–1156 (2011).
103. Li, W.-S.; Yamamoto, Y.; Fukushima, T.; Saeki, A.; Seki, S.; Tagawa, S.; Masunaga, H.; Sasaki, S.; Takata, M. and Aida, T. Amphiphilic molecular design as a rational strategy for tailoring bicontinuous electron donor and acceptor arrays: photoconductive liquid crystalline oligothiophene-C60 dyads, *J. Am. Chem. Soc.*, **130**, 8886–8887 (2008); DOI: 10.1021/ja802757w.

104. Park, H. J.; Kang, M.-G.; Ahn, S. H. and Guo, L. J., A facile route to polymer solar cells with optimum morphology readily applicable to a roll-to-roll process without sacrificing high device performance, *Adv. Mater.*, **22**, E247–E253 (2010); DOI: 10.1002/adma.201000250.
105. Getachew, A.; Almantas, P.; Alberto, M. R.; Sisay, T.; Teketel, Y.; Niyazi, S. S. and Daniel, A. M. E., Mobility and photovoltaic performance studies on polymer blends: effects of side chains volume fraction, *J. Mater. Chem.*, **21**, 2594–2600 (2011); DOI: 10.1039/c0jm02668a.
106. Kimber, R. G. E.; Walker, A. B.; Schröder-Turk, G. E. and Cleaver, D. J., Bicontinuous minimal surface nanostructures for polymer blend solar cells, *Phys. Chem. Chem. Phys.*, **12**, 844–851 (2010); DOI: 10.1039/b916340a.
107. Raffaelle, R. P., Photovoltaics research & development at the U. S. National Renewable Energy Lab, *Nature Photonics Technology Conference*, October 2010 19–21, Tokyo, Japan; http://www.natureasia.com/en/events/photonics/pdf/c-5.pdf.
108. Andreev, V. M.; Grilikhes, V. A.; Khvostikov, V. P.; Khvostikova, O. A.; Rumyantsev, V. D.; Sadchikov, N. A. and Shvarts, M. Z., Concentrator PV modules and solar cells for TPV systems, *Solar Energy Mater. Solar Cells*, **84**(1–4), 3–17 (2004).
109. Janssen, A. G. F.; Riedl, T.; Hamwi, S.; Johannes, H.-H. and Kowalsky, W., Highly efficient organic tandem solar cells using an improved connecting architecture, *Appl. Phys. Lett.*, **91**, 073519 (2007); DOI: 10.1063/1.2772208.
110. Mayer, A.; Scully, S.; Hardin, B.; Rowell, M. and McGehee, M., Polymer-based solar cells, *Mater. Today*, **10**, 28–33 (2007); doi:10.1016/S1369–7021(07)70276–6.
111. Hoppe, H. and Sariciftci, N. S., Polymer solar cells, in *Photoresponsive Polymers II*, Marder, S. R. and Lee, K.-S., Eds., *Adv Polym Sci.*, **214**, 1–86, Springer, ISBN 978–3-540–69452–6, Berlin-Heidelberg (2008).
112. Na, S.-I.; Kim, S.-S.; Jo, J. and Kim, D.-Y., Efficient and flexible ITO-free organic solar cells using highly conductive polymer anodes, *Adv. Mater.*, **20**(21), 4061–4067 (2008); DOI: 10.1002/adma.200800338.
113. Richards, B. S. and Shalav, A., Photovoltaic devices, in *The Handbook of Photonics*, 2nd Edn., Ballato, J. and Gupta, M. C., Eds., CRC Press (2007); ISBN: 978–0-8493–3095–7.
114. Ardalan, P.; Brennan, T. P.; Lee, H.-B.-R.; Bakke, J. R Ding, I/-K.; McGehee, M. D. and Bent, S. F., Effects of self-assembled monolayers on solid-state CdS quantum dot sensitized solar cells, ACSNANO, 5(2), 1495–1504 (2011).
115. Lee, S. H.; Xu, J.; Zhang, X.-G.; Parish, C. M. and Smith, B., Nanojunction solar cells based on polycrystalline CdTe films grown on ZnO nanocones, *IEEE Proceedings* (2011); Xu, J.; Lee, S. H.; Zhang, X.-G.; Parish, C. M.; Smith, B.; Duty, C. and Lee, H. N., Efficient charge transport in nanocone tip-film solar cells, *IEEE Proceedings 37th IEEE Photovoltaic Specialist Conference*, 19–24 June 2011, Washington State Convention Center, Seattle, WA (2011). http://www.ornl.gov/info/press_releases/get_press_release.cfm?ReleaseNumber=mr20110429–00.
116. Gao, F.; Wang, Y.; Shi, D.; Zhang, J.; Wang, M. K.; Jing, X. Y.; Humphry-Baker, R.; Wang, P.; Zakeeruddin, S. M. and Gratzel, M. Enhance the optical absorptivity of nanocrystalline TiO2 film with high molar extinction coefficient ruthenium sensitizers for high performance dye-sensitized solar cells, *J. Am. Chem. Soc.*, **130**, 10720–10728 (2008).

117. Gutmann, F. and Lawrence, L., *Organic Semiconductors: Part A*, Krieger Pub Co., Melbourne, FA, USA (August 1983); ISBN-10: 0882758233.
118. Bruetting, W., *Physics of Organic Semiconductors*, Wiley-VCH, Hoboken, NJ, USA (2005); ISBN-10: 352740550X.
119. Chamberlain G. A., Organic solar cells: a review, *Solar Cells*, **8** 47–83 47 (1983).
120. Chopra, K. L.; Paulson, P. D. and Dutta, V., Thin-film solar cells: An overview, *Prog. Photovolt: Res. Appl.*, **12**, 69–92 (2004); DOI: 10.1002/pip.541.
121. Hoppe, H. and Sariciftci, N. S., Morphology of polymer/fullerene bulk heterojunction solar cells, *J. Mater. Chem.*, **16**, 45–61 (2006); DOI: 10.1039/b510618b.
122. Günes, S.; Neugebauer, H. and Sariciftci, N. S., Conjugated polymer-based organic solar cells, *Chem. Rev.*, **107**, 1324–1338 (2007); DOI: 10.1021/cr050149z.
123. Liang, Y.; Xu, Z.; Xia, J.; Tsai, S.-T.; Wu, Y.; Li, G.; Ray, C. and Yu, L., For the bright future–Bulk heterojunction polymer solar cells with power conversion efficiency of 7.4%, *Adv. Mater.*, **22**, E135–E138 (2010).
124. OPV cell achieves record 8.3 percent efficiency, PV Magazine, 30 November 2010; http://www.pv-magazine.com/news/details/beitrag/.
125. Wu, Y.; Currier, T.; Li, Y. and Tsai, S.-T., Organic light emitting diode and organic solar cell stack, WO 2011057191 (A1), 2011–05–12, to Solarmer Energy Inc [USA].
126. Troshin, P. A.; Hoppe, H.; Peregudov, A. S.; Egginger, M.; Shokhovets, S.; Gobsch, G.; Sariciftci, N. S. and Razumov, V. F., [70]Fullerene-based materials for organic solar cells, *ChemSusChem*, **4**, 119–124 (2011); DOI: 10.1002/cssc.201000246.
127. Khlyabich, P. P.; Burkhart, B.; Ng, C. F. and C. Thompson, B. C., Efficient solar cells from semi-random P3HT analogues incorporating diketopyrrolopyrrole, *Macromolecules*, **44**, 5079–5084 (2011); DOI.org/10.1021/ma2009386.
128. Huo, L.; Guo, X.; Zhang, S.; Li, Y. and Hou, J., PBDTTTZ: A broad band gap conjugated polymer with high photovoltaic performance in polymer solar cells, *Macromolecules*, **44**, 4035–4037 (2011); DOI.org/10.1021/ma200743b.
129. Tajima, K. and Hashimoto, K., Nanostructure control in polymer solar cells by self-organization, *Chem. Record*, **11**(1), 8–17 (2011); DOPI 10.1002/tcr.201000015.
130. Bower, D. I., *An Introduction to Polymer Physics*, Cambridge University Press, Cambridge, UK (2002); ISBN-10 0–521–63137–8.
131. IEC-904–3 Measurement Principles for Terrestrial PV Solar Devices with Reference Spectral Irradiance Data.
132. Benanti, T. L. and Venkataraman, D., Organic solar cells: An overview focusing on active layer morphology, *Photosynth. Res.*, **87**, 73–81 (2006); DOI: 10.1007/s11120–005–6397–6399.
133. Kroon, J. M.; Wienk, M. M.; Verhees, W. J. H. and Hummelen, J.C. Accurate efficiency determination and stability studies of conjugated polymer / fullerene solar cells, *Thin Solid Films*, **2002**, 403–404, 223–228 (2002).
134. Ma, W.; Yang, C.; Gong, X.; Lee, K. and Heeger, A., Thermally stable, efficient polymer solar cells with nanoscale control of the interpenetrating network morphology, *Adv. Funct. Mater.*, **15**, 1617–1622 (2005); DOI: 10.1002/adfm.200500211.
135. Riisager, L. B. L., Polymer solar cells – Steps towards improving the power conversion efficiency, Master's Thesis, Aalborg University, Denmark (2009).
136. Wu, K., Advanced materials and fabrication methods for organic solar cell, Open Access Dissertations and Theses. Paper 4140 (2010); http://digitalcommons.mcmaster.ca/opendissertations/4140.

137. Gilot, J.; Wienk, M. M. and Janssena, R. A. J., Double and triple junction polymer solar cells processed from solution, *Appl. Phys. Lett.*, **90**, 143512 (2007); DOI: 10.1063/1.2719668.
138. Dastoor, P. C.; McNeill, C. R.; Frohne, H.; Foster, C. J.; Dean, B.; Fell, C. J.; Belcher, W. J.; Campbell, W. M.; Officer, D. L.; Blake, I. M.; Thordarson, P.; Crossley, M. J. and Hush, N. S.; Reimers, J. R., Understanding and improving solid-state polymer/C60-fullerene bulk-heterojunction solar cells using ternary porphyrin blends, *J. Phys. Chem. C*, **111**, 15415–15426 (2007); DOI: 10.1021/jp0748664.
139. Helgesen, M.; Gevorgyan, S. A.; Krebs, F. C. and Janssen, R. A. J., Substituted 2,1,3-benzothiadiazole- and thiophene-based polymers for solar cells – introducing a new thermo-cleavable precursor, *Chem. Mater.*, **21**, 4669–4675 (2009); DOI:10.1021/cm901937d.
140. Breeze, A. J.; Schlesinger, Z.; Carter, S. A.; Tillmann, H. and Hörhold, H.-H., Improving power efficiencies in polymer–polymer blend photovoltaics, *Solar Energy Materials & Solar Cells*, **83**, 263–271 (2004); DOI:10.1016/j.solmat.2004.02.029.
141. Scharber, M. C.; Mühlbacher, D.; Koppe, M.; Denk, P.; Waldauf, C.; Heeger, A. J. and Brabec, C. J., Design rules for donors in bulk-heterojunction solar cells–towards 10% energy-conversion efficiency, *Adv. Mater.*, **18**, 789–794 (2006); DOI: 10.1002/adma.200501717.
142. Peet, J.; Heeger, A. J.; Bazan, G. C., Plastic solar cells: self-assembly of bulk heterojunction nanomaterials by spontaneous phase separation, *Acc. Chem. Res.*, **42**(11) 1700–1708 (2009); DOI: 10.1021/ar900065j.
143. Li, G.; Shrotriya, V.; Huang, J.; Yao, Y.; Moriarty, T.; Emery, K. and Yang, Y., High-efficiency solution processable polymer photovoltaic cells by self-organization of polymer blends, *Nature Mater.*, **4**, 864–868 (2005); DOI: 10.1038/nmat1500.
144. Chen, H.-Y.; Hou, J.; Zhang,S.; Liang, Y.; Yang, G.; Yang, Y.; Yu, L.; Wu, Y. and Li, G., Polymer solar cells with enhanced open-circuit voltage and efficiency, *Nature Photonics*, **3**, 649–653 (2009); DOI: 10.1038/NPHOTON.2009.192.
145. Ihn, S.-G.; Shin, K.-S.; Jin, M.-J.; Bulliard, X.; Yun, S.; Choi, Y. S.; Kim, Y.; Park, J.-H.; Sim, M.; Kim, M.; Cho, K.; Kim, T. S.; Choi, D.; Choi, J.-Y.; Choi, W. and Kim, S.-W., ITO-free inverted polymer solar cells using a GZO cathode modified by ZnO, *Solar Energy Materials & Solar Cells*, **95**, 1610–1614 (2011); DOI:10.1016/j.solmat.2011.01.011.
146. Yang, X.; van Duren, J. K. J.; Janssen, R. A. J.; Michels, M. A. J. and Loos, J., Morphology and thermal stability of the active layer in poly(p-phenylenevinylene)/methanofullerene plastic photovoltaic devices, *Macromolecules*, **37**, 2151–2158 (2004); DOI: 10.1021/ma035620.
147. Changsarn, S.; Mendez, J. D.; Weder, C. and Supaphol, P., Morphology and photophysical properties of electrospun light-emitting polystyrene/poly- (p-phenylene ethynylene) fibers, *Macromol. Mater. Eng.*, **293**, 952–963 (2008); DOI: 10.1002/mame.200800268.
148. Bradley, M. and Izzia, F., ATR Spectra of materials from the near to far-infrared, The Application Notebook (2004); www.thermo.com/ftir.
149. He, Y.; Chen, H.-Y.; Zhao, G.; Hou, J. and Li., Y., Synthesis and photovoltaic properties of biindene-C70 monoadduct as acceptor in polymer solar cells, *Solar Energy Materials & Solar Cells*, **95**, 1762–1766 (2011); DOI: 10.1016/j.solmat.2011.01.045.

150. Verduzco, R.; Botiz, I.; Pickel, D. L.; Kilbey II, S. M.; Hong, K.; Dimasi, E. and Darling, S. B., Polythiophene-block-polyfluorene and polythiophene-block-poly(fluorene-co-benzothiadiazole): Insights into the self-assembly of all-conjugated block copolymers, *Macromolecules*, **44**, 530–539 (2011); DOI: 10.1021/ma102728z.
151. Woody, K. B.; Leever, B. J.; Durstock, M. F. and Collard, D. M., Synthesis and characterization of fully conjugated donor–acceptor–donor triblock copolymers, *Macromolecules*, **44**(12), 4690–4698 (2011); DOI: 10.1021/ma200084z.
152. Woodhouse, M.; Perkins, C. L.; Rawls, M. T.; Cormier, R. A.; Liang, Z.; Nardes, A. M. and Gregg, B. A., Non-conjugated polymers for organic photovoltaics: physical and optoelectronic properties of poly(perylene diimides), *J. Phys. Chem. C*, **114**, 6784–6790 (2010); DOI: 10.1021/jp910738a.
153. Utracki, L. A., *Clay-Containing Polymeric Nanocomposites*, 786 pp. RAPRA, Shawbury, Shrewsbury, Shropshire UK (2004).
154. Kittichungchit, V.; Hori, T.; Moritou, H.; Kubo, H.; Fujii, A. and Ozaki, M., Effect of solvent vapor treatment on photovoltaic properties of conducting polymer/C60 interpenetrating heterojunction structured organic solar cell, *Thin Solid Films*, **518**(2), 518–521 (2009).
155. United States Air Force Systems Command, Aerospace materials, Volumes 1–7 of AFSC design handbook, The Command (1987).
156. Buhl, H., *Advanced Aerospace Materials*, Springer-Verlag, New York (1992).
157. Cantor, B.; Grant, P. and Assender, H., Eds., Aerospace materials (Graduate student series in materials science and engineering), Institute of Physics Publishing, Dirac House, Bristol, UK (2001).
158. http://standards.sae.org/materials/polymers/standards/.
159. Solar Impulse: Soaring to New Heights in Solar Aviation; http://suprememastertv.com/gat/?wr_id=91.
160. Ross, H., Around the world with a solar powered aircraft, http://www.mp.haw-hamburg.de/pers/Scholz/dglr/hh/text_2011_01_20_SolarImpuls.pdf.
161. http://www.solvayplastics.com/sites/solvayplastics/EN/Companies/ Solvay_Advanced_Polymers/Sustainability/Solvay_AP_Solar_Impulse-r1.pdf.
162. IKAROS, http://www.jspec.jaxa.jp/ikaros_cam/e/.
163. JAXA (Japan Aerospace Exploration Agency) Press release of 23 July 2010, Small Solar Power Sail Demonstrator 'IKAROS' Successful Attitude Control by Liquid Crystal Device.
164. Edwards, L., IKAROS unfurls first ever solar sail in space, 11.06.2010; http://www.physorg.com/news195460006.html.
165. Iwata, M.; Suzuki, M.; Kazama, K.; Ando, A.; Miyauchi, M.; Ishida, Y. and Yokota, R., Properties of newly developed thermoplastic polyimide and its durability under proton, electron, and UV irradiations, *Proceedings of the 11th International Symposium on Materials in Space Environment*, 15–18.09.2009, Aix-en-Provence, France.
166. Smith, Jr., J. G.; Smith, T.; Williams, M.; Youngquist, R. and Mendell, W., Potential polymeric sphere construction materials for a spacecraft electrostatic shield, National Aeronautics and Space Administration Langley Research Center, Hampton, VA 23681, NASA/TM–2006–214302.
167. Nair, C. P. R.; Mathew, D. and Ninan, K. N., Cyanate ester resins, recent developments, *Adv. Polym. Sci.*, **155**, 1–99 (2001).

168. Carter, J.T.; Emmerson, G.T.; Faro, C. Lo; McGrail, P.T. and Moore, D.R., The development of a low temperature cure modified epoxy resin system for aerospace composites, *Composites: Part A*, **80**, 83–91 (2003).
169. Gavrin, A.; Nebo, J.; Rice, N.; Kagumba, L.; Smith, D. and Jin, J.; Topping, Ch., Perfluorocyclobutane (PFCB) polyaryl ethers for space-based applications, Protection of Materials and Structures from Space Environment, Space Technology Proceedings, 2004, **5**(5), 491–501, DOI: 10.1007/1–4020–2595–5_45.
170. Bhowmik, S.; Bonin, H. W.; Bui, V. T. and Weir, R. D., Modification of high-performance polymer composite through high-energy radiation and low-pressure plasma for aerospace and space applications, *J. Appl. Polym. Sci.*, **102**, 1959–1967 (2006).
171. Miyazaki, E.; Tagawa, M.; Yokota, K.; Yokota, R.; Kimoto, Y. and Ishizawa, J., Investigation into tolerance of polysiloxane-block-polyimide film against atomic oxygen, *Acta Astronautica*, **66**(5–6) 922–928 (2010).
172. Malmongea, L. F.; do Carmo Langianoa, S.; Marques Cordeiroa, J. M.; Capparelli Mattosob, L. H. and Malmongea, J. A., Thermal and mechanical properties of PVDF/PANI blends, *Mater. Res.*, **13**(4), 465–470 (2010).
173. Hoffmann, T.; Pospiech, D.; Häussler, L.; Pötschke, P.; Reuter, U.; Werner, P.; Sandler, J. K.W.; Döring, M. and Altstädt, V., Properties of segmented block copolymers in PEEK/PSU blends, *High Perform Polym.*, **20**, 601–614 (2008); DOI: 10.1177/0954008307082176.
174. Zhou, J.; Li, A.; Huang, H. and Rao, B., Polyimide-epoxy blend for binding silicon steel sheets of aerospace electrical machineries, *Key Eng. Mater.*, **353–358**, 3055–3058 (2007); DOI:10.4028/www.scientific.net/KEM.353–358.3055.
175. Theiler, G. and Gradt, T., MoS2-filled PEEK composite as a self-lubricating material for aerospace applications, *Proceedings of the 40th Aerospace Mechanisms Symposium*, NASA Kennedy Space Center, 12–14 May 2010, NASA/CP-2010–216272; pp. 347–351.
176. De Fenzo, A.; Formicola, C.; Antonucci, V.; Zarrelli, M. and Giordano, M., Effects of zinc-based flame retardants on the degradation behaviour of an aerospace epoxy matrix, *Polym. Degrad. Stab.*, **94**(9), 1354–1363 (2009); DOI:10.1016/j.polymdegradstab.2009.05.020.
177. Tighe, A. P.; van Eesbeek M. R. J. and Wernham, D., Spacecraft materials test techniques utilising high power lasers, *Proceedings of the 10th ISMSE & the 8th ICPMSE, Collioure*, France, 19–23 June 2006.
178. Banks, B. A.; Waters, D. L.; Thorson, S. D.; de Groh, K. K., Snyder, A. and Miller, S., Comparison of atomic oxygen erosion yields of materials at various energies and impact angles, *Proceedings of the 10th ISMSE & the 8th ICPMSE, Collioure*, France, 19–23 June 2006; de Groh, K. K.; Banks, B. A.; McCarthy, C. E.; Rucker, R. N.; Roberts, L. M. and Berger, L. A., Analysis of the MISSE PEACE polymers international space station environmental exposure experiment, *Proceedings of the 10th ISMSE & the 8th ICPMSE*, Collioure, France, 19–23 June 2006; Dever, J. A.; Miller, S. K. and Sechkar, E. A., Effects of the space environment on polymer film materials exposed on the materials International Space Station experiment (MISSE 1 and MISSE 2), *Proceedings of the 10th ISMSE & the 8th ICPMSE*, Collioure, France, 19–23 June 2006; Banks, B. A.; Backus, J. A.; Manno, M. V.; Waters, D. L.; Cameron, K. C. and de Groh, K. K., Atomic oxygen erosion yield prediction for spacecraft polymers in low earth

orbit, http://esmat.esa.int/Materials_News/ISME09/pdf/7-Databases/S9%20-%20Banks.pdf.

179. de Groh, K. K. and Banks, B. A., MISSE 2 PEACE polymers erosion morphology studies, *11th ISMSE International Symposium on Materials in a Space Environment*, 15–18 September 2009, Aix-en-Provence, France.
180. Fuse, N.; Tanaka, T. and Ohki, Y. Evaluation of dielectric properties in polypropylene/clay nanocomposites, *2009 Ann. Rep. IEEE Conf. Electr. Insul. Dielectr. Phenom.*, Virginia Beach, VA, 18–21 October 2009.
181. Utracki, L. A., Clay-containing polymeric nanocomposites and their dielectric properties (Feature Article), *IEEE Electrical Insulation Magazine*, **26**(4), 6–15 (2010); art. no. 5511184.
182. Morgan, A. B. and Wilkie, C. A., Eds., *Flame Retardant Polymer Nanocomposites*, Wiley, Hoboken, NJ (2007); ISBN: 978–0-471–73426–0.
183. Tagawa, M. and Minton, T. K., Mechanistic studies of atomic oxygen reactions with polymers and combined effects with vacuum ultraviolet light, *MRS Bulletin*, **35**(1), 35–40 (2010).
184. Zhang, J.; Lindholm, N. F.; Brunsvold, A. L.; Upadhyaya, H. P. and Minton, T. K., Studies of the individual and combined effects of VUV radiation and hyperthermal O or Ar atoms on FEP Teflon and PMMA surfaces, Protection of materials and structures from space environment. *Proceedings of the 9th International Conference*, pp. 154–169, Kleinman, J. I., ed., *Am Inst. Phys.* (2009).
185. Yuan, H.; Killelea, D. R.; Tepavcevic, S.; Kelber, S. I. and Sibener, S. J., interfacial chemistry of poly(methyl methacrylate) arising from exposure to vacuum-ultraviolet light and atomic oxygen, *J. Phys. Chem. A*, **115**, 3736–3745 (2011); DOI.org/10.1021/jp1061368.
186. Shimamura, H. and Miyazaki, E., Investigation into synergistic effects of Atomic Oxygen and Vacuum Ultraviolet, *J. Spacecraft Rockets*, **46**(2), 241–247 (2009).
187. Lévy, L.; Dirassen, B.; Reulet, R.; Van Eesbeek, M. and Molinié, P., Dark and radiation induced conductivity on space used external coatings, *Proceedings of the 10th ISMSE & the 8th ICPMSE*, Collioure, France, 19–23 June 2006.
188. Bucknall, C. B. and Gilbert, A. H., Toughening tetrafunctional epoxy resins using polyetheramide, *Polymer*, **30**, 213–217 (1989).
189. Carfagna, C.; Nicolais, L.; Amendola, E.; C. Carfagna, Jr., C. and Fllippov, A. G., Toughening epoxy resins by liquid crystalline polymers, *J. Appl. Polym. Sci.*, **44**, 1465–1471 (1992).
190. Weil, E. D., New anti-corrosive coatings with resin-bonded polyaniline and related elecytroactive groups, Final report for grant No. NAG10–0174, April 1997.
191. International Trade Bridge (ITB) Inc., National Aeronautics Space Administration (NASA) acquisition pollution prevention (AP2) office, Contract No. NAS1–03029, Joint test protocol for validation of alternatives to aliphatic isocyanate polyurethanes, Final NAP2.Proj.JTP.Aiu.Pl.01.31.05.F, 31 January 2005.
192. International Trade Bridge (ITB) Inc., National Aeronautics Space Administration (NASA) acquisition pollution prevention (AP2) office, Contract No. NAS1–03029, Joint test protocol for validation of alternatives to aliphatic isocyanate polyurethanes, Final NAP2.Proj.JTP.Aiu.Pl.01.31.05.F, 19 May 2006.
193. Helm, M. L.; Stewart, M. P.; Bullock, R. M.; Rakowski-DuBois M. and DuBois, D. L., A synthetic nickel electrocatalyst with a turnover frequency above 100,000 s–1 for H2 production, *Science*, **333**, 863–866 (2011).

194. Duncan, A. J.; Leo, D. J. and Long, T. E., Beyond Nafion: charged macromolecules tailored for performance as ionic polymer transducers, *Macromolecules*, **41** (21) 7765–7775 (2008); DOI: 10.1021/ma800956v.
195. Gubler, L. and Scherer, G. G., Trends for fuel cell membrane development, *Desalination*, **250**, 1034–1037 (2010); DOI:10.1016/j.desal.2009.09.101.
196. Matras-Postolek, K. and Bogdal, D., Polymer nanocomposites for electro-optics: perspectives on processing technologies, material characterization, and future application, *Adv. Polym. Sci.*, **230**, 221–282 (2010); DOI: 10.1007/12_2010_49.
197. Hindson, J. C.; Saghi, Z.; Hernandez-Garrido, J.-C.; Midgley, P. A. and Greenham, N. C., Morphological study of nanoparticle-polymer solar cells using high-angle annular dark-field electron tomography, *Nano Lett.*, **11**, 904–909 (2011), DOI.org/10.1021/nl104436j/
198. Ridolfi, G.; Casalbore-Miceli, G.; Geri, A.; Camaioni, N., Possamai, G. and Maggini, M., Organic photovoltaics: Towards a revolution in the solar industry, *Proceedings of A Consultants Meeting, Bologna,* Italy, 22–25 March 2004, *Emerging applications of radiation in nanotechnology,* IAEA-TECDOC-1438, pp.71–83 (2004).
199. Yuan, Q. and Misra, R. D. K., Polymer nanocomposites: current understanding and issues, *Mater. Sci. Technol.*, **22**, 742–755 (2006); DOI: 10.1179/174328406X101292.
200. Utracki, L. A.; Sepehr, M. and Boccaleri, E., Synthetic, layered nano-particles for polymeric nanocomposites (PNC's), *Polym. Adv. Technol.*, **18(1)**, 1–37 (2007); DOI: 10.1002/pat.852.
201. Kim, H.; Abdala, A. A. and Macosko, C. W., Graphene/polymer nanocomposites, *Macromolecules*, **43** (16), 6515–6530 (2010); DOI: 10.1021/ma100572e.

3.11 Appendix: Nomenclature

3.11.1 General abbreviations

AA	acrylic acid (monomer)
AF	amorphous fluoro
A-FC	alkaline fuel cell
AFM	atomic force microscopy
AM	acoustic microscopy
AMTA	American Membrane Technology Association
AO	atomic oxygen
aq.	aqueous
ATR	attenuated total reflection (e.g., in FT-IR-ATR)
BDT	benzodithiophene
BHJ	bulk heterojunction
CA	cellulose acetate
CE	counter electrode
CIGS	copper-indium-gallium *di*-selenide, $Cu(In,Ga)Se_2$

CUT	continuous use temperature
CV	cyclic voltammetry
DM-FC	direct methanol fuel cell
DSC	differential scanning calorimetry
DSSC	dye-sensitized solar cell
DTA	differential thermal analysis
EIS	electrochemical impedance spectroscopy
EPFL	École Polytechnique Fédérale de Lausanne
EQE	external quantum efficiency
ESCA	electron spectroscopy for chemical analysis
EVS	electrochemical voltage spectroscopy
FC	fuel cell
FEM	field emission microscopy
FF	fill factor
FVMQ	silicone rubber with fluorine, vinyl, and methyl substituents
GC	gas chromatography
GC-MS	gas chromatography-mass spectrometry
GPC	gel permeation chromatograph (now: size exclusion chromatography, SEC)
HOMO	highest occupied molecular orbital
HP-FC	hydrogen-powered fuel cell
HTP	high temperature polymer
HTPB	high temperature polymer blend
IEC	ion-exchange capacity
IQE	internal quantum efficiency
ISS	International Space Station
ITO	indium tin oxide (transparent hole collector in SC)
JAXA	Japan Aerospace Agency
LCD	liquid crystal device for attitude control by light reflection
LCP	liquid crystal polymer
LUMO	lowest unoccupied molecular orbital
MALLS	multi-angle laser light scattering
MAS-NMR	magic angle spinning NMR
MC-FC	molten carbonate fuel cell
MCT	mercury cadmium telluride detector
MD	molecular dynamics calculation
NF	nanofiltration
NGDC-FM	natural gas direct carbon fuel cell
NIR	near-infrared spectroscopy
NMR	nuclear magnetic resonance
NREL	US National Renewable Energy Laboratory

OM	optical microscopy
OOP	onset of oxidation potential
OSC	organic solar cell
PA-FC	phosphoric acid fuel cell
PALS	positron annihilation lifetime spectroscopy
$PC_{60}BM$	6,6-phenyl C_{61}-butyric acid methyl ester (also: PCBM)
$PC_{70}BM$	6,6-phenyl C_{71}-butyric acid methyl ester
$PC_{70}PP$	phenyl-C_{71}-propionic acid propyl ester
$PC_{70}PB$	phenyl-C_{71}-propionic acid butyl ester
PCE	power conversion efficiency
PDE	partial discharge erosion
PEM-FC	polymer exchange membrane fuel cell
PEST	thermoplastic polyesters, e.g., PBT, PET, also TPES
PPV	*p*-phenylene vinylene
PSC	polymeric solar cell
PV	photovoltaic
QD	quantum dot
RH	relative humidity (in %)
RI	refractive index
RO	reverse osmosis
RT	room temperature
SAE	Society of Aerospace and Automotive Engineers
SAED	selected-area electron diffraction
SAXS	small-angle X-ray scattering
SC	solar cell
SEC	size exclusion chromatography
SEM	scanning electron microscopy
SO-FC	solid oxide fuel cell
SRP	self-reinforced polyphenylene ether
STC	standard test conditions
SWNT	single-wall carbon nanotube
TEM	transmission electron microscopy
TGA	thermogravimetric analysis
TMS	tri-methylsilyl, or tetra-methyl silane
TOR	*trans*-polyoctenamer rubber
TPI	thermoplastic polyimide
TTZ	thiazolo[5,4-d]thiazole
UF	ultrafiltration
VFTH	Vogel-Fulcher-Tamman-Hesse equation
VUV	vacuum ultra violet
wt%	weight percent
WU	water uptake (e.g., by membrane)
XPS	X-ray photoelectron spectroscopy

XRD	X-ray diffraction
XSBR	butadiene/styrene/acidic monomer; an elastomeric copolymer

3.11.2 Polymer abbreviations

ABS	thermoplastic terpolymer, an acrylonitrile butadiene styrene copolymer
BCP	block copolymer
CAB	cellulose acetate butyrate
CF	cresol-formaldehyde resins (also reinforcing carbon fiber)
CR	chloroprene, or neoprene, rubber
DAD	donor-acceptor-donor
ECTFE	poly(ethylene-co-chloro tri-fluoro ethylene)
EPDM	elastomeric terpolymer from ethylene, propylene, and a non-conjugated diene
EPR	elastomeric copolymer of ethylene and propylene
ETFE	copolymer from ethylene and tetrafluoroethylene
F8BT	poly(9,9′-dioctylfluorene-*co*-benzothiadiazole)
FEP	fluorinated EPR; tetrafluoroethylene/hexa-fluoro propylene rubber
FKM	hexa-fluoro propylene-vinylidenefluoride copolymer
FVMQ	silicone rubber with fluorine, vinyl, and methyl substituents
NBR	elastomeric copolymer from butadiene and acrylonitrile; nitrile rubber
PA	polyamide; the abbreviation PA is normally followed by a number, a combination of numbers, a letter or a combination of letters and numbers. A single number refers to the polyamide from an α,ω-amino acid or its lactam. A combination of two numbers is often separated by a comma. The first number following the symbol PA indicates the number of methylene groups of aliphatic di-amines, the second number the number of carbon atoms of aliphatic di-carboxylic acids. An I stands for isophthalic acid, a T for terephthalic acid. For example, co-polyamide from caprolactam, hexamethylenediamine condensed with isophthalic and terephthalic acids is abbreviated as: PA-6IT6, or that from caprolactam, *m*-xylylenediamine and adipic acid as: PA-mXD6, etc.
PAB	polymer alloys and blends
PACE	polyacetylene
PAEK	polyaryletherketone

PAES	polyarylethersulfone
PAI	polyamide-imide
PAN	polyacrylonitrile
PANI	polyaniline
PAr, PAR	polyarylate, $[-\phi\text{-}C(CH_3)_2\text{-}\phi\text{-}CO_2\text{-}\phi\text{-}CO_2\text{-}]_n$, amorphous polyester of bisphenol-A with isophthalic and terephthalic acids
PARA	polyaryl amide (aromatic, usually amorphous polyamide)
PBD	polybutadiene
PBI	polybenzimidazole, e.g., poly[2,2′-(*m*-phenylene)-5,5′*bis*-benzimidazol]
PBO	polybutyleneoxide
PBT,	PBTP polybutyleneterephthalate
$PC_{60}BM$	6,6-phenyl C_{61}-butyric acid methyl ester (also: PCBM)
PCTG	copolymer of cyclohexanedimethanol (≥ 66 mol %), ethylene glycol (≤ 34 mol %) and terephthalic acid
PCTFE	polychlorotrifluoroethylene
PDMS	polydimethylsiloxane
PEA	polyetheramide
PEBA	thermoplastic elastomer, polyether-block-amide
PEC	polyestercarbonate or chlorinated polyethylene; usually CPE
PEDOT	poly(3,4-ethylenedioxythiophene)
PEEK	polyetheretherketone
PEI	polyetherimide
PEK	polyetherketone
PEM-FC	polymer exchange membrane fuel cell
PEN	poly(ethylene 2,6-naphthalene dicarboxylate) or polyethylenenaphthalate
PES	polyethersulfone $[-\phi\text{-}SO_2\text{-}\phi\text{-}O\text{-}]_n$
PEST	thermoplastic polyesters, e.g., PBT, PET, also TPES
PEU	polyether urethane
PFA	polyfluoroalcoxyalkane; copolymer of tetrafluoroethylene and perfluorinated
PFB	poly(9,9′-dioctylfluorene-*co*-bis-*N,N′*-(4-butyl-phenyl)-bis-*N,N′*-phenyl-1,4-phenylenediamine)
PG	poly-α-hydroxy acrylic acid
PI	polyimide, but also *trans*-1,4-polyisoprene, gutta-percha (UK)
PMB	poly*p*-methylenebenzoate
PMDMO-PPV	poly[2-methyl,5-(3,7-dimethyloctyloxy dimethyl-octyloxy)-*p*-phenylene vinylene]
PP	polypropylene

PPA	polyphthalamide
PPD-T, PPTA	poly(*p*-phenylene terephthalamide), *Kevlar*™
PPhA	polyphthalamide
PPS	polyphenylsulfide
PPSS, PPS-S	polyphenylene sulfidesulfone, polythioethersulfone
PPSU, PSF,	PSO polyphenylene sulfone; polysulfone
PPTA, PPD-T	poly(1,4-phenylene terephthalamide)
PS	polystyrene
PS-b-PVBPA	poly(styrene-b-vinyl-benzyl phosphonic acid)
PSF	polysulfone, also PSUL, PSU, PSO
PSO	polysulfone, also PSUL, PSU, PSF
PSS	poly(styrene-sulfonate)
PSU	polysulfone $\left[\text{-}\phi\text{-SO}_2\text{-}\phi\text{-O-}\phi\text{-C(CH}_3)_2\text{-}\phi\text{-O-}\right]_n$
PU, PUR	polyurethane elastomer
PVAc	polyvinyl acetate
PVAl	polyvinyl alcohol
PVDC	polyvinylidene chloride; also PVC2
PVDF	polyvinylidene fluoride; also PVF2
PVF	polyvinyl fluoride
PVP	poly-N-vinylpyrrolidone
SPEKK	sulfonated poly(ether ketone ketone)
SPPS	sulfonated polyphenylene sulfide
TPI	thermoplastic polyimide

3.11.3 Notation

B_{ij}	Huggins-Flory type binary interaction parameters between segments (i and j)
d_d	limit of phase heterogeneity in PAB
D_f	dissipation factor
D_k	dielectric constant
E_a	Arrhenius activation energy
FF	fill factor
I	current
I_{sc}	min(I_{sci}) short circuit voltage
M_n	number average molecular weight
M_v	viscosity-average molecular weight
M_w	weight average molecular weight
P	pressure
P_e	electrical power generated by the PV cell
P_{in}	incidental power of solar radiation
q	heating rate

Q	irradiation intensity
R	gas constant
T	temperature
T1 NMR	spin-lattice relaxation times
T_b	boiling point
T_c	crossover temperature
$T_{c,m}$	middle of crystallization temperature
$T_{c,o}$	onset of crystallization temperature
T_d	degradation temperature
T_g	glass transition temperature
t_{gel}	gel time
T_m	melting temperature
T_{pyr}	pyrolysis temperature is proportional
V	volume or voltage
V_{oc}	open circuit voltage

3.11.4 Greek

ϕ	volume fraction
ΔE	activation energy
ΔG_m	free energy of mixing
ΔH_c	heat of crystallization
ΔH_m	heat of melting
ΔH_m	heat of mixing
Δl	interphase thickness in PAB
ΔT_g	width of T_g peak at half-height
δ	Hildebrand solubility parameter
$\gamma = \omega / B_0$	nuclei gyro-magnetic ratio of Larmor frequency to magnetic field strength
η	viscosity
η_r	relative viscosity
η_{inh}	inherent viscosity
$\eta_{PCE} = I_{sc} V_{oc} FF / P_{in}$	power conversion efficiency (also PCE)
$[\eta]$	intrinsic viscosity
λ	wavelength
σ	conductivity in polymer electrolytes
$\sigma = n\mu$	proton conductivity, product of the number of charges (n) per unit volume and the ion mobility in an electric field (μ)

4
Thermodynamics of high temperature polymer blends

M. T. DEMEUSE, MTD Polymer Consulting, USA

DOI: 10.1533/9780857099013.130

Abstract: The performance and subsequent properties of polymer blends are highly dependent on the blend's phase structure. For example, a miscible mixture of two polymers will have different features to an immiscible mixture of the same two polymers. Additionally, the manner in which a transformation from a miscible blend to an immiscible blend occurs will affect the ultimate properties. These features can be categorized under the topic of thermodynamics of polymer blends. This chapter discusses some features of the thermodynamics of blends that contain high temperature polymers, highlighting those which are most important in defining the blend phase structure. A comparison is made with other polymer blends, and important differences are noted.

Key words: thermodynamics, miscible, Flory–Huggins, glass transition, liquid crystal polymer.

4.1 Introduction

The performance of polymer blends depends on the properties of the polymeric components, as well as on how they are arranged in space. The spatial arrangement is controlled to a large extent by the thermodynamics of the system. The term 'thermodynamics' invariably brings to mind 'miscibility'. The aim of this chapter is to discuss the thermodynamic features of high temperature polymer blends and highlight studies that have focused on a determination of the features that lead to miscibility in such systems.

A miscible polymer blend is a homogeneous, single phase material. Several experimental methods, many of which have already been discussed in Chapter 2, have been developed for defining such a structure in a mixture of polymers. Among these methods, the occurrence of a single glass transition temperature, or T_g, is the most widely accepted criteria for miscibility. It is generally assumed that polymer blends that display a single T_g are molecularly mixed. Based on those criteria, various models[1,2] have been developed in the polymer literature for a description of the relation of a blend T_g and the blend composition.

A necessary condition for two polymers to form a miscible mixture is that the free energy of mixing, ΔG_{mix}, should be less than zero. In general, ΔG_{mix} is given by the following equation:

$$\Delta G_{mix} = \Delta H_{mix} - T\Delta S_{mix} \quad [4.1]$$

where ΔH_{mix} is the enthalpy of mixing, T is the temperature and ΔS_{mix} is the entropy of mixing. Obviously, one way for ΔG_{mix} to be negative is for ΔH_{mix} to be negative. This is generally accomplished through specific interactions between the two polymers in the mixture, such as dipole–dipole interactions or hydrogen bonding. However, simply having a negative free energy of mixing does not guarantee phase stability. The necessary condition for phase stability in a blend composition of x at a fixed temperature, T, and pressure, P, is that the second derivative of the free energy with respect to composition is greater than zero, i.e.

$$\frac{\mathrm{d}\Delta G_{mix}}{\mathrm{d}x^2} > 0 \quad [4.2]$$

Without the satisfaction of this condition, the miscible system is thermodynamically unstable and phase separation will occur.

One of the most common models for dealing with the free energy of mixing between two polymers is the Flory–Huggins model.[3] That model assumes that the free energy of mixing, ΔG_{mix}, for the two polymers is given by:

$$\frac{\Delta G_{mix}}{RT} = \frac{\Theta_1}{N_1}\ln\Theta_1 + \frac{\Theta_2}{N_2}\ln\Theta_2 + \chi_{12}\Theta_1\Theta_2 \quad [4.3]$$

where Θ_1 and Θ_2 are the volume fractions, respectively, of polymers 1 and 2, N_1 and N_2 are the degrees of polymerization of polymers 1 and 2 and χ is the so-called 'chi' parameter used to describe the enthalpic interaction between polymers 1 and 2. This model, as originally developed, is strictly applicable only to polymers that have random coil configurations. It is also an assumption that the entropic and enthalpic contributions to the free energy are completely independent of each other.

Flory[4,5] first proposed the concept of molecular composites which are systems based on the mixing of a rigid-rod polymer and a random coil polymer, vastly different molecular conformations. The theoretical prediction was made that phase separation is easily induced in such systems. The phase separation in such blend systems is based solely on entropic effects. This is an important difference with random coil mixtures in which the driving force for phase separation is usually non-favorable enthalpic interactions.

Two extreme cases of behavior have, then, been theoretically examined in the literature. Mixing of two random coil polymers is controlled by enthalpic effects while the mixing of rigid-rod polymers with random coils is dominated by entropic effects. Since most high temperature polymers contain some form of heterocyclic units, they have restricted rotation in their backbone. It is exactly that restricted motion that often leads to the high glass transition temperature observed in high temperature polymers. Thus, most high temperature polymers can be treated as rigid, or at least semi-rigid, in overall conformation. An adequate theoretical model is still to be developed for the mixing of such polymers that have a molecular conformation between random coils and rigid rods.

It should be noted that there have been attempts to produce molecular modeling results and simulations of high temperature polymer blends. The most extensive of these efforts was published by Jacobson *et al.*[6] They used a simple short chain molecular model that incorporates both inter and intramolecular interactions. Using that model, estimates of the net interaction energies for a series of high temperature polymer blends were calculated and used to predict miscibility. The results were in general agreement with experiments and could be used to focus the direction of additional experimental work.

4.2 Blending miscible high temperature polymers

There have been several miscible high temperature polymer pairs that have been identified in both patent and open literature. Several of these polymer pairs are miscible when solutions are formed and films are cast. However, when processing is attempted in the melt state, immiscibility results. These results suggest that the blends phase separate when heated above their glass transition temperature. Also, kinetic factors along with thermodynamic factors seem to be affecting the observed miscibility.

The role of the solvent itself also needs to be better understood in the overall scheme of the miscibility observed. It may be that the presence of the solvent produces a kinetically favorable situation for miscibility but, upon attainment of thermodynamic equilibrium, that situation no longer exists. One of the challenges for many of the systems that will be discussed is to broaden the temperature range between the glass transition temperature of the initially miscible blend and its phase separation temperature. Such an effect will allow miscible blends to be processed in the melt state, thus eliminating the need for processing and handling of solvents in the fabrication of products from the blends.

4.3 Poly (2,2' (*m*-phenylene)-5-5' bibenzimidazole) (PBI) blends

One of the most highly studied high temperature polymers in miscible blends is poly (2,2′ (*m*-phenylene)-5-5′ bibenzimidazole) or PBI, the chemical structure of which is shown in Fig. 4.1. The fundamental reason for the observation of miscibility in PBI-based systems is the presence of the N-H functional group that can interact with the functional groups which are present on the backbone of other polymers. Thus, miscibility in these types of systems is an example of a specific interaction that leads to a negative enthalpy of mixing.

The most extensively studied PBI blend system is PBI blended with a polyetherimide, originally available from GE, Ultem 1000.[7] The chemical structure of Ultem is shown in Fig. 4.2. The early work on that blend led to the phase separation diagram shown in Fig. 4.3. This figure shows that a processing window exists for blends that contain in excess of 75% PBI and also for blends that contain less than about 20% PBI. In this case, the processing window is defined as those temperatures between T_g and the phase separation temperature. For example, for an 85/15 PBI/Ultem blend ratio, the glass transition temperature is 400°C and the processing window is 25°C.

Infrared studies of these blends showed that hydrogen bonding exists between the N-H groups of PBI and the carbonyl group of Ultem[8] and that the hydrogen bonding relaxes during the thermal treatment phase separation.[9,10] In subsequent work,[11] a direct relationship was found between the strength of the hydrogen bonding of the component polymers, the glass

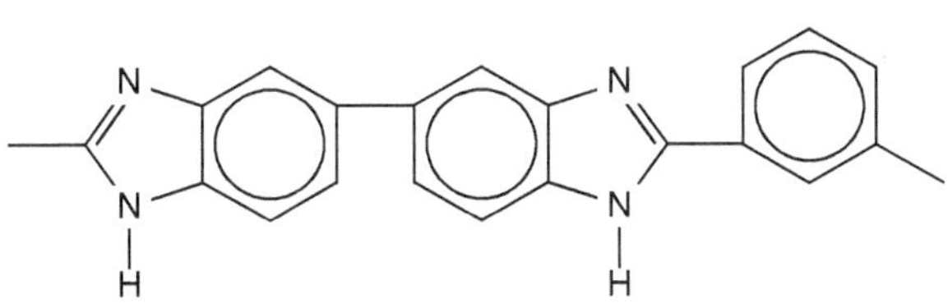

4.1 Chemical structure of PBI polymer.

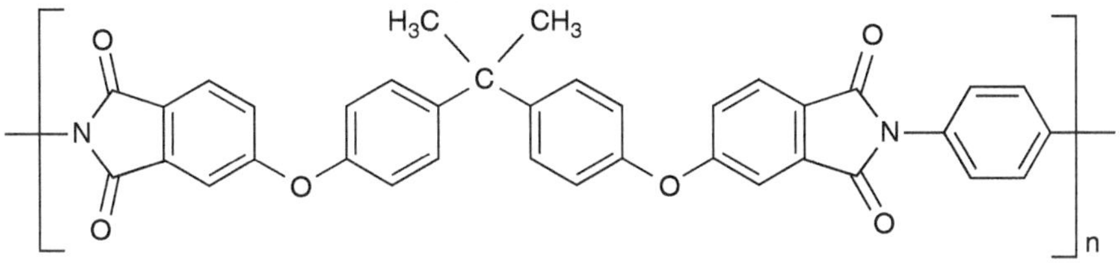

4.2 Chemical structure of Ultem polyetherimide.

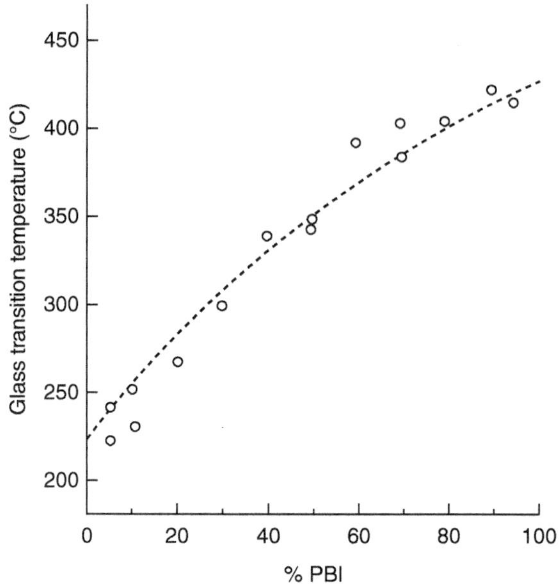

4.3 Phase separation diagram for PBI/Ultem blends.

transition temperature of the blends and the solvent diffusion rate of both water and 1,2,4 trichlorobenzene. These latter studies conclude that there is a partial miscibility between the two polymers and that phase inversion occurs between 0.4 and 0.6 Ultem content.

The phase behavior observed with the PBI/Ultem blends is to be contrasted with the phase separation observed in blends of PBI with another polyimide, XU218 from Ciba Geigy.[12] That system showed phase separation only above 400°C. The actual temperatures are determined by the blend composition. Thus, the phase behavior observed in PBI/Polyimide blends is dependent both on the type of polyimide and the thermal history of the blends.

In further investigations of PBI blends, Jaffe *et al.*[13] showed that PBI is miscible with certain polyimides that contain the hexafluoroisopropylidene or 6F moiety, i.e. CF_3–C–CF_3. The miscibility and phase behavior in these blend systems is dependent on the overall structure of the polyimide. In addition, the amount of 6F chemical group affects the T_g of the polyimide itself and, hence, the T_g of any miscible blends.

Those studies of blends of PBI with various polyimides were subsequently extended to include other polymers. For example, it was shown that PBI and polysulfone form immiscible mixtures.[14] However, it was later shown[15] that the introduction of functional groups, such as sulfonate groups, into the polysulfone polymer chain resulted in the formation of miscible blends with

PBI. It was shown that the sulfonation degree as well as the blend composition controls the miscibility behavior. FT-IR analysis confirmed that the observed miscibility is due to specific interactions between the PBI N-H groups and the sulfonate groups on the polysulfone.

4.4 Polyimide blends

There are other high temperature polymers that have also been shown to form miscible mixtures. Polyimides (PIs) have already been discussed in blends with PBI. Other miscible PI blends have been reported with poly (ether-ether ketone) (PEEK),[16] polyethersulone (PES)[17] and sulfonated PEEK.[18] Although the mechanism for miscibility in PBI/PIs was demonstrated to be related to hydrogen bonding between the chains of the two components, the mechanism of miscibility in some of these other blend systems was not as clear.

In order to better understand the miscibility in polyimide-based systems, Sun *et al.*[19] prepared pairs of polyimide blends with different molecular structures in two ways: mixing of the polyamic acid precursors with subsequent imidization and direct solution blending of the two polyimides. Dynamic mechanical analysis (DMA) techniques showed that all of the blends prepared in the two different ways are miscible, as evidenced by the existence of only one Tg for all the blends. It was proposed that the miscibility of these polyimide/polyimide blends is a result of the strong intermolecular charge-transfer interaction between the two chains of the blend components.

Overall, then, the high temperature polymer blends that have been discussed thus far display some type of well-defined specific interaction that leads to the observed miscibility. In the case of PBI blends, it is hydrogen bonding that occurs through the N-H group present on the PBI backbone. For PI blends, there appears to be a charge-transfer interaction that leads to miscibility in mixtures. In both cases, it is clearly the enthalpic part of the free energy of mixing that leads to miscible blends.

4.5 Liquid crystal polymer blends

Such is not the case in the blends of two liquid crystal polymers (LCPs) that were first extensively studied by DeMeuse and Jaffe. In their initial study,[20] they examined blends of LCPs that contain copolymers of p-hyrdoxybenzoic acid (HBA) and 6-hydroxynapthoic acid (HNA) of different copolymer ratios. It was surmised that the miscibility observed in the melt state depended on the difference in copolymer ratios between the two component polymers. However, miscibility in the solid state seemed to be present for all the blends examined. This is an interesting contrast to the systems mentioned above because no obvious enthalpic interaction occurs between the two blend components.

In further studies[21,22] those preliminary efforts were extended to include LCPs that contained other monomer units such as terephthalic acid (TA) and hydroquinone (HQ). In some cases, miscible systems were defined[21] and sometimes immiscible blends were observed.[22] For all cases, however, there was not a well-defined specific interaction that could be attached to the observed miscibility. It seems that the entropy of mixing is important in defining the miscibility in these types of systems.

An extensive study of a number of blends that contain only thermotropic liquid crystalline polymers as components was undertaken by Hsieh *et al.*[23] A wide selection of commercial or semi-commercial materials was chosen for that study. Both thermotropic LCPs (TLCPs) that are regarded as rigid and semi-rigid in molecular conformation were studied. The study revealed that TLCPs with few common comonomers and/or with different rigidities are generally found to be immiscible. Polymeric aspects such as molecular chemistry and molecular packing are decisive aspects. For miscibility to occur, both components should be of similar molecular rigidity, rigid or semiflexible, and preferably contain some common monomer or monomers.

Thus, we have two different types of behavior in blends of high temperature polymers. The first situation is exemplified by the PBI blends in which specific interactions through the N-H group of the PBI are responsible for the observed miscibility. The other situation is displayed by the mixtures of two LCPs in which entropic effects are important for the miscibility. These two types of behavior would seem to be the extreme cases. This also suggests that there are blend systems of high temperature polymers in which both enthalpic and entropic effects should be important for miscibility.

This concept was explored by Lee and DiBenedetto[24] who introduced a second LCP as a compatibilizing agent in order to improve the adhesion and dispersion between the components of incompatible LCP/thermoplastic blends. The primary reason that the LCP and thermoplastic polymer are immiscible is due to molecular conformation differences or entropy effects, as originally discussed by Flory in his work on molecular composites. The concept for using a second LCP in such blends is that the two LCPs will be miscible due to entropy effects and the second LCP and the thermoplastic polymer will adhere due to specific interactions. The LCP coupling phase used in this work was a copolymer of PET and HBA known as PHB60. Blends of an LCP with both polycarbonate (PC) and PET were prepared with and without the addition of the second LCP coupling phase. Morphological evidence indicated that the LCP reinforcing phase in the ternary systems exhibited improved adhesion and dispersion on a finer scale than in the binary blends that were prepared using the same processing conditions.

This approach was further explored by Hakemi[25] who prepared blends that contain both a wholly aromatic and an aromatic-aliphatic LCP that are miscible with each other. The ultimate goal of this approach was to develop multi-component miscible blends that have components of thermoplastics. The miscible blends could be useful as reinforcing agents for the thermoplastic matrix polymer, and because LCPs contain some of the components of the thermoplastic polymer, improved adhesion between the LCP portion and the matrix portion of the mixture is expected. This is another example of an attempt to balance the phase separation inherent in high temperature polymer blends due to molecular conformation differences by enforcing the enthalpic interactions between the two polymers.

Before leaving the topic of LCP/LCP blends, one additional point needs to be stressed. Low molecular weight liquid crystals (LMWLCs) of the same mesophasic class often show miscibility.[26] On the other hand, TLCPs that form the same mesophase, usually nematic, do not necessarily exhibit complete miscibility, Thus, in order to implement the ideas of the last several paragraphs, care must be used in choosing the two LCPs to form the miscible blend.

4.6 Molecular composites

As already discussed in the Introduction (Section 4.1), molecular composites, as first envisioned by Flory, show phase separation based strictly on entropic effects. Those entropic effects arise from the differences in molecular conformation between the rigid rod and the random coil component in the mixture. Thus, in order to interrupt the phase separation, one possible approach is to enhance the potential for specific interactions through the incorporation of interacting chemical groups into the polymeric components of the mixture.

Initially, ionic blends were utilized to produce miscibility in molecular composites.[27–30] Ionic bonds are stronger and more thermally stable than hydrogen bonds and, thus, are deemed to be more effective at promoting miscibility. Those initial studies focused primarily on molecular composites that were cast from solution. After those initial studies, work was also focused on the development of melt-processable molecular composites. Such mixtures were produced by dispersing rigid-rod molecules, such as ionic versions of Kevlar, poly (p-phenylene terephthalamide) (PPTA), in a matrix of a flexible polymer, such as poly (4-vinylphenol) (PVP).[31–33] Relatively low additions of a PPTA anion, which contains the ionic groups directly attached to the backbone chains, in the PVP matrix, led to miscibility and to a good dispersion of the rigid rods in the matrix. These miscibility effects were attributed to the presence of ionic interactions between the ionic groups of the modified PPTA and the polar groups of the vinyl pyridine units.

This work represents another attempt at overcoming the phase separation inherent in mixtures of polymers of vastly different molecular conformations through the use of specific molecular interactions. That seems to be an effective approach for interrupting the phase separation. The general application of this idea to the development of specific high temperature polymer blends is still to be realized, however. Also, it needs to be better understood how the chemical modification of the rigid-rod component to increase the molecular interactions alters its conformation, if at all, toward more of a semi-rigid polymer. This could also contribute to the enhanced miscibility with other high temperature polymers.

4.7 Conclusions

In general, the production of miscible blends of high temperature polymers is driven by the need to have specific molecular interactions that lead to favorable values of the enthalpy of mixing portion of the free energy of mixing. While this is often also the case with mixtures of flexible coil polymers, the miscibility criteria for mixtures of high temperature polymers seem to be more stringent. As one moves away from random coil/random coil mixing, entropic effects, other than simply molecular weight, seem to play an increasingly important role. This can be seen clearly from the fact that mixtures of true rigid rods with random coils are immiscible based solely on entropic effects that arise from differences in molecular conformations.

Practically speaking, however, very few polymers can be categorized as rigid rod in conformation, but instead fall into the category of semi-rigid in molecular conformation. In fact, most high temperature polymers that have been blended can best be classified as semi-rigid. Therefore, the theories that have been developed to define the thermodynamics of mixtures of polymers do not apply to these polymers.

This is an area for which additional work is recommended. Establishment of a theory or model that explicitly captures the conformation of the two polymers being mixed is highly desired. Presently, extreme cases have been defined, and the transition from one extreme to the other needs to be better understood. Also, the ability of a polymer to change its molecular conformation to maximize possible favorable enthalpic interactions such as hydrogen bonding should be considered in the theoretical development.

From an experimental perspective, it becomes clear that miscible blends offer certain advantages. Further, an obvious way to enhance the possibility of miscibility in mixtures that involve high temperature polymers is through the development of specific interactions. That approach should be further explored with high temperature polymers through the incorporation of chemical groups that can promote such interactions. A caution about that general approach is that the appropriate chemical modification does not cause the polymer to completely lose its high temperature features.

4.8 Sources of further information and advice

There do not appear to be any books devoted solely to high temperature polymer blends, let alone the thermodynamics of such mixtures. A reasonable review article of some of the factors that affect the phase behavior of polymer blends for high temperature applications was provided by Jaffe *et al.*[34]

4.9 References

1. M. Gordon and J.S. Taylor, *J. Appl. Chem* **2** (1952) 493.
2. T.K. Kwei, *J. Polym. Sci. Lett.* **22** (1984) 307.
3. P.J. Flory, *J. Chem. Phys.* **10** (1942) 51–61.
4. A. Abe and P.J. Flory, *Macromolecules* **11** (1978) 1122.
5. P.J. Flory, *Macromolecules* **11** (1978) 1138.
6. S.H. Jacobson, D.J. Gordon, G.V. Nelson and A. Balasz, *Advanced Materials*, Vol. 4 (3) (1992), 198–205.
7. T. Takekoshi *et al.*, *J. Polym. Sci., Polym. Chem. Ed.* **18** (1980) 3069.
8. G. Guerraa, S. Choe, D.J. Williams, F.E. Karasz and W.J. MacKnight, *Macromolecules* **21** (1988) 231.
9. P. Musto, F.E. Karasz and W.J. MacKnight, *Macromolecules* **24** (1991) 4762.
10. S. Choe, F.E. Karasz and W.J. MacKnight, in B.M. Culbertson, editor, *Contemporary Topics in Polymer Science, Multiphase Macromolecular Systems* Vol. 6, New York, Plenum Press, (1989) 493.
11. E. Foldes, E. Fekete, F.E. Karasz and B. Pukanszky, *Polymer* **41** (2000) 975–983.
12. S. Choe, W.J. MacKnight and F.E. Karasz in C. Feger, M.M. Khojaseth and J.E. McGrath, editors, *Polyimides: Materials, Chemistry and Characterization*, Amsterdam, Elsevier, (1989) 25.
13. M. Jaffe *et al.*, *Polym. Eng. Sci.* **32**(17) (1992) 1235–1241.
14. T.S. Chung, M. Glick and E. Powers, *Polym. Eng. Sci.* **33**(16) (1993) 1042–1048.
15. V. Deimede, G.H. Voyiatzis, J.K. Kullitsis, L. Qingfeng and N.J. Bjerrum, *Macromolecules* **33**(24) (2000) 7609–7617.
16. Y. Kong, H. Tang, L. Dong, F. Teng and Z Feng, *J. Polym. Sci. Part B: Polymer Physics* **36**(13) (1998) 2267–2274.
17. K. Liang, J. Grabowicz, E. Valles, F.E. Karasz and W.J. MacKnight, *J.Polym. Sci., Part B: Polymer Physics* **30**(5) (1992) 465–476.
18. R.J. Karcha and R.S. Porter, *J. Polym. Sci., Part B: Polym. Phys.* **27** (1989) 2153.
19. Z. Sun, H. Li, Y. Zhang, M. Ding and Z. Feng, *Polym. Bull.* **26** (1991) 557–563.
20. M.T. DeMeuse and M. Jaffe, *Molec. Cryst. Liq. Cryst.*, **157** (1988) 535.
21. M.T. DeMeuse and M. Jaffe, *Polym. Prepr.* **30** (1989) 540.
22. M.T. DeMeuse and M. Jaffe, in R.A. Weiss and C.K. Ober, editors, *Liquid Crystalline Polymers*, ACS Symposium Series No. 435 (1990) 439.
23. T.T. Hsieh, G. Tiu and G.P. Simon, *Polymer* **41** (2000) 4737–4742.
24. C.W. Lee and A.T. DiBenedetto, *Polym. Eng. Sci.* **32** (1992) 400.

25. H. Hakemi, *Polymer* **41** (2000) 6145–6150.
26. G.W. Gray and P.A. Windsor in *Liquid Crystals and Plastic Crystals* Vol. 1, Halsted Press, New York (1974) 20.
27. G. Parker, W. Chen. L. Tsou and M. Hara, *ACS Symp. Ser.* **632** (1996) 54.
28. G. Parker, W.Chen and M. Hara, *Polym. Mater. Sci. Eng.* **72** (1996) 534.
29. C.D. Eisenbach, J. Hoffmann and W.J. MacKnight, *Macromolecules* **27** (1994) 3162.
30. R.A. Weiss, L. Shao and R.D. Lundberg, *Macromolecules* **25** (1992) 6370.
31. M. Hara and G. Parker, *Polymer* **33** (1992) 4650.
32. G. Parker and M. Hara, *Polymer* **38** (1997) 2701.
33. G. Parker and M. Hara, *Polymer* **38** (1997) 2773.
34. M. Jaffe, P. Chen, E.C. Chenevey, T.S. Chung and S. Makhija, in *High Performance Polymers: Advances in Polymer Science* Vol. 117 (1994) 297–327.

5
Liquid crystal polymers (LCPs) as a reinforcement in high temperature polymer blends

M. T. DeMEUSE, MTD Polymer Consulting, USA and
G. KISS, Telecordia Technologies, USA

DOI: 10.1533/9780857099013.141

Abstract: Liquid crystal polymers (LCPs) are a relatively new class of high temperature polymers that have attracted a great deal of interest, both academically and industrially. This interest has arisen because of the very high mechanical properties that LCPs provide, as well as the relatively low melt viscosity. There have also been attempts to take advantage of these unique properties in mixtures with other polymers. This chapter will summarize those studies and highlight the advantages offered by the use of LCPs in blends, including an effect referred to as a self-reinforcing composite. It is also possible to use LCPs as rheology modifying agents. Future directions in the area are recommended.

Key words: liquid crystal polymer, thermoplastic, blend, self-reinforcing, immiscible.

5.1 Introduction

Liquid crystal polymers (LCPs) are a relatively new class of high temperature polymers. Blends of these materials have been produced that take advantage of the unique features of the LCPs, including the very high physical properties they offer. This chapter will summarize the activities, both academic and industrial, that have been carried out in this area. The idea of creating a reinforcement species *in situ* while conducting processing operations of a material which is unreinforced prior to processing is an attractive and elegant way to sidestep problems encountered with conventional filled materials, such as increased viscosity, degradation of the reinforcing species, contamination of the final product and wear on the tooling. This concept is not limited to polymer blends, but occurs in metallurgy as well.

In the field of polymer engineering, several approaches have been considered. One is to exploit the fact that extended-chain crystals of polymers are both more stable (higher melting point) and stronger (not weakened by the presence of chain folding). We will not discuss the extensive literature on solution-grown extended-chain crystals nor that of solid-state extrusion, but confine ourselves to thermoplastic processing. An early example is by Janssen *et al.*[1] in which the possibility of forming a self-reinforcing composite of a single material (polypropylene) by extrusion under conditions such that extended-chain reinforcing fibers are formed within a matrix of chain-folded material. Another approach to this problem is that of *in situ* polymerization in which a monomer species is created within a matrix such that a post-processing polymerization step can occur, leading to the eventual polymeric reinforcing species. Examples of this are Collyer[2] and Kiss *et al.*[3]

The focus of this chapter is more specifically blends in which the reinforcing species is a thermotropic LCP which is blended in a rather conventional way with isotropic thermoplastics (TP). This exploits the inherent strength and low viscosity of the LCP in a way which leads to maximum benefit at minimum processing complexity. This has been an actively studied area of polymer engineering for 40 years and continues to be of interest. We note that though much work has been done on the science of LCP blends, there is also a need to review the exploitation aspects in terms of cost considerations, patents and commercial products.

5.2 Researching liquid crystal polymers (LCPs)

Before the advent of the internet and easily accessible online search tools, there was great value in a work which purported to be a complete and exhaustive overview of a specific aspect of a field. Presumably the author would have read essentially every relevant article from the major journals in the field, as well as a selection of obscure journals, from bound volumes and current issues in a library. This was done as a service to others with an interest in the field, to spare fellow researchers the effort of having to hunt down all of the primary materials themselves. The production of review and survey articles was part of the process of digesting new science and incorporating it into the body of established science. The unspoken assertion of the author of such a work was that 'if it is not in my survey/review article, it is not significant'.

Exhaustiveness as a virtue in its own right has been made obsolete and trivial by online search and research tools, which allow any given article to be the entrée to both an infinite regress and an infinite progress of citations. This chapter can be seen as something of a sampler, in which many references are included as a starting point for locating further related work, rather than the final reference. It is truly an awe-inspiring experience for one who was an active researcher in the pre-internet days, when 'literature search' was synonymous with many days in the library, many hours at the copy machine, and many pages of handwritten notes, to experience contemporary online search, such as Web of Science.[4] (One of us [GK] would like to express heartfelt appreciation to Lisa Klein of Rutgers for guest access to Web of Science.) In real time one can start with a search such as '*in situ* composite' and find 175 hits, and see them immediately or export them to a convenient list of abstracts for perusal. Each hit is characterized by a very useful piece of metadata: the number of times the work has been cited in more recent work. This allows one to see immediately which papers have been heavily cited and are 'required reading' in the field.

Alternatively, one can start with the 175 '*in situ* composite' hits, then add additional terms such as 'polymer' (26 hits), 'LCP' (12 hits) and 'rheology' (5 hits) and focus rapidly and efficiently onto the niche of interest, and have the papers (and their reference lists) in hand immediately as PDFs on a USB stick. Thus many days of pre-internet literature search can be accomplished in a single working session. The term continues to be used to describe LCP/thermoplastic polymer blends, as does the term *self-reinforcing composite* (though the latter is a misnomer, since the matrix does not in fact reinforce itself; there is a distinct reinforcing species).[5,6]

The references are divided into three categories. First are the traditional references to scientific and trade journals. Second, we have chosen to segregate the patents which are a component of the technical literature to be sure, but not motivated primarily by a desire to disseminate information and instruct, but by a desire to exploit the technology commercially. Many of the questions which naturally occur to the reader are left deliberately obscure in patent writing. Third, we include a category for 'links and personal communication' which expands on the ephemeral conversations and letters which had traditionally been referenced to include online resources in the form of links which may or may not be viable for future readers. Online search makes it simple to track down persons who might be willing to confirm or discuss a 'personal communication'. Also, the internet archive (aka the 'Wayback Machine') makes even expired links potentially viable.

Although we have said that review and survey articles are less critical now than in the pre-internet days, nevertheless they represent the best resource for information about discoveries and progress as this field emerged in the 1960s, 1970s and 1980s. It continued to generate substantial interest in the 1990s, by which time online resources were becoming available. In the 2000s and to the present day, LCP blends continue to generate interest both in journal articles and patents, and in conferences.

An indispensable reference which reviewed the state of the art in 1990 is Dutta, Fruitwala, Kohli, and Weiss.[7] The rheology of LCPs was reviewed by Wissbrun.[8] A more recent review is from 1996,[9] particularly Chapter 4 by Cogswell and Wissbrunn[10] and Chapter 6 by Collyer.[11] Three early papers published in a single issue of Polymer Engineering and Science V27 (1987) have each been cited hundreds of times as of August 2011.[12–14] These papers popularized the term *in situ composite* to indicate the notion that the reinforcing species does not exist per se in the unformed material, but rather comes into existence during processing.

5.3 Liquid crystals

Much of the material in this section can be seen in an online webinar, courtesy of Malvern Instruments.[15] Liquid crystals, also known as ordered fluids, are an intermediate state of matter between ordinary liquids and crystalline solids. Like ordinary liquids they flow in response to stress, although some exhibit yield stress. Like crystalline solids they exhibit some degree of long range order, in some cases forming elaborate patterns which extend for distances many orders of magnitude greater than the molecular length. The subject of liquid crystals is vast and has been studied intensely for many decades, the earliest citation being from 1888. The most familiar liquid crystal phenomena are optical effects exploited for displays, and involve low molecular weight compounds.[16,17]

Liquid crystals can be formed by melting of crystalline solids, in which case they are called thermotropic, or by dissolving certain solids in a solvent, in which case they are called lyotropic. Both of these routes to liquid crystals can give rise to a panoply of structures (called 'mesophases'), of which the best known are nematic, cholesteric, and smectic.

The simplest of the structures is the nematic, which is formed by molecules with high aspect ratio ('rod-like') and characterized by a degree of long range order in the orientation of the long axis of the molecules but not in any other geometric feature. A schematic view of a small-molecule nematic such as the extensively studied para-azoxyanisole (p-azoxyanisole), is shown in Fig. 5.1a and 5.1b. Most small-molecule nematics can be heated sufficiently that they lose their orientational order before decomposing, forming a conventional isotropic liquid as shown in Fig. 5.1c. This temperature is often called the 'clearing point' because there is usually a loss of turbidity.

The nematic-to-isotropic transition also occurs for lyotropic small-molecule liquid crystals. Looking at Fig. 5.1b and 5.1c, one can easily imagine that the oriented molecules would slide past one another with relative ease in the nematic. In isotropic melts the molecules at rest have no preferred orientation and would interfere with each other to a greater extent when sheared. Also, they would be expected to display shear-thinning as shear-induced orientation is created. Thus one would expect a lower zero-shear viscosity in the nematic melt, although it exists at a lower temperature than the isotropic melt. This is indeed the case, and a sudden increase in viscosity at the clearing temperature is seen in many nematics as the orientational

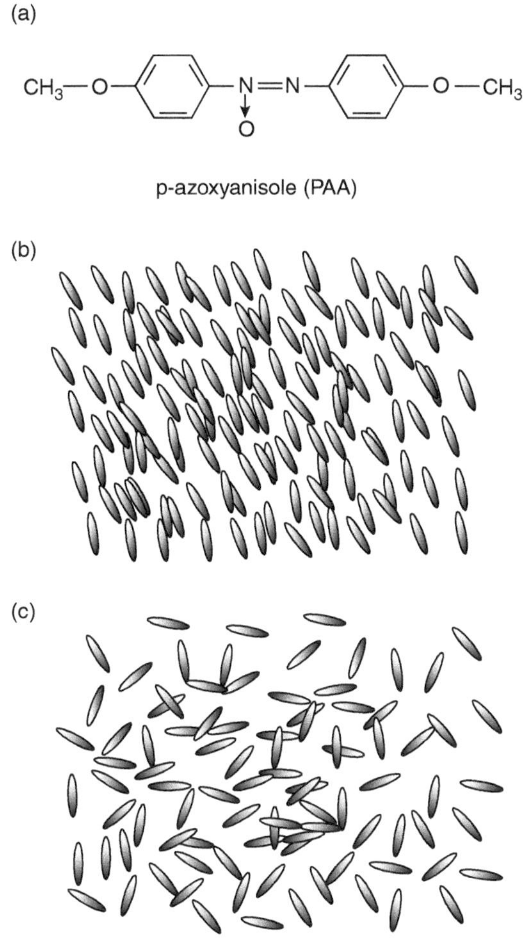

5.1 (a) Chemical structure of p-azoxyanisole molecule. (b) Nematic structure of p-azoxyanisole. (c) Isotropic liquid structure of p-zoxyanisole.

order at rest disappears (see Fig. 5.2). The rheology of small-molecule liquid crystals was reviewed in 1962.[18]

5.4 Polymer liquid crystals

Flexible polymers generally do not form mesophases; however, polymers which have a high aspect ratio are sometimes able to. There are two broad categories of rod-like polymers:

- Those which have a rigid backbone, composed entirely of rigid bonds, which happen to be arranged so that the overall molecule is elongated.

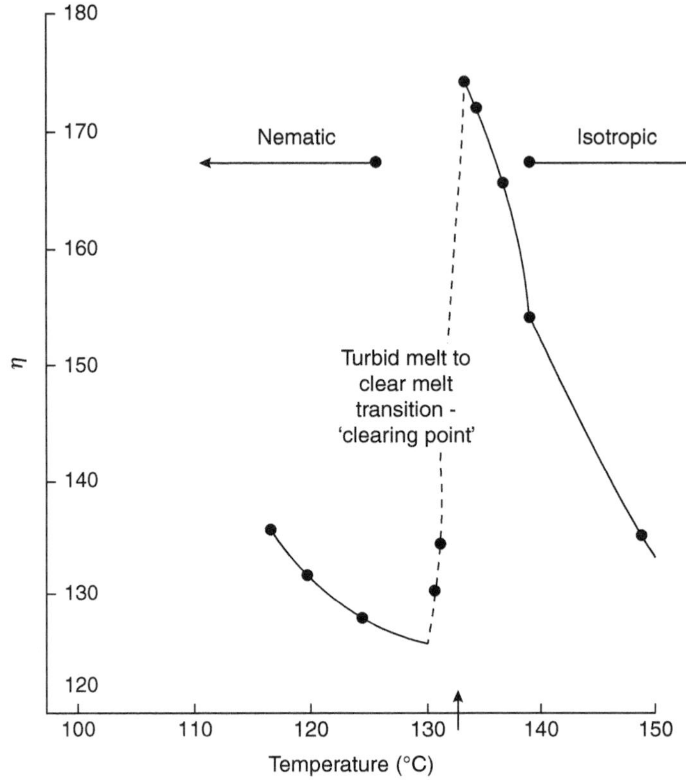

5.2 Transition temperatures of p-azoxyanisole.

Not incidentally, polymers with extended backbones have high modulus, since elongating the bulk polymer requires bonds to be stretched, rather than bent.

- Those which have a flexible backbone which happens to take on a conformational structure such that the overall molecule is elongated. The best-known example of this is helical molecules such as DNA and PBLG, but in fact much early work was done using a living rod-like polymer, namely tobacco mosaic virus.[19,20]

Examples exist of both types of rod-like polymers forming both lyotropic and thermotropic mesophases. An exploration of the correspondence between the two types is given in reference 21.

5.4.1 Lyotropic polymer solutions

Rigid backbone polymers are typically difficult to dissolve. A significant application of lyotropic polymer solutions is the nematic spinning dope which is obtained when dissolving para-aromatic polyamides ('aramids') in powerful solvents such as concentrated sulfuric acid. Note that the polymer itself is not liquid crystalline, as commonly stated, but rather the spinning dope is a nematic. The best-known example is DuPont's Kevlar, and the commercial significance of Kevlar inspired a great deal of activity around both lyotropic and thermotropic polymer liquid crystals, starting in the 1960s.[22–24]

The viscosity behavior seen in small-molecule nematics is echoed in lyotropic polymer solutions. Concentration plays a role analogous to that of temperature. Just as increasing temperature causes the loss of zero-shear orientation in small-molecule nematics, so too does decreasing concentration. When concentration falls below a critical value, which is related to aspect ratio, polymer solute molecules are able to assume random orientation due to the availability of sufficient volume per molecule[25], and a phenomenon analogous to the viscosity peak at the clearing point is seen. That is, a peak in zero-shear viscosity vs. concentration is observed with the analogous surprising observation that the more concentrated (anisotropic) solutions have lower viscosity than the less concentrated (isotropic) solutions of the popular model system poly-gamma-benzyl-glutamate in m-cresol.[26,27] Furthermore, the effect of the volume-filling orientation is seen to be most pronounced when extrapolated to zero-shear, and it decreases in magnitude while shifting to lower concentration as shear rate is increased (see Fig. 5.3), and eventually disappears entirely when the shear-induced orientation overwhelms the volume-filling effect.

The viscosity behavior is thus simple, straightforward and unsurprising. Not so the normal stress behavior. The same model system gives bizarre and unexpected normal stress behavior in that steady-state values of normal stress are observed at well-defined shear rates intermediate between a low shear rate positive negative transition and a high shear rate negative positive transition, as seen in Fig. 5.4.[23] This behavior is well documented, and a satisfactory theoretical model has been developed.[28]

5.4.2 Thermotropic liquid crystal polymers

Thermotropic liquid crystal polymers are very analogous to small-molecule nematics. They are (partially) crystalline solids which melt to form anisotropic melts. As stated earlier, polymers which are rigid and rod like typically have melting points higher than their decomposition temperatures, so must be processed from solution. This obviously restricts the shapes available to fibers and films. Much effort has been expended to create rod-like polymers which can melt sufficiently to be molded and yet retain the desired mechanical properties. Various approaches to slightly disrupting the rod packing have been tried. (Griffin and Cox[21] referred to this process using the vivid

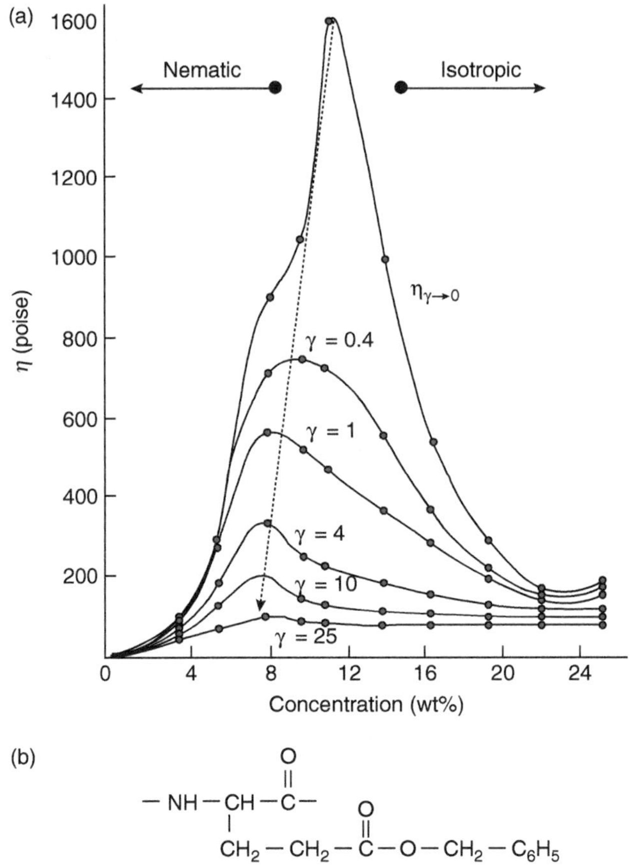

5.3 (a) Viscosity at various shear rates vs. concentration for 335 000 MW PBLG in m-cresol and (b) chemical structure of poly-gamma benzyl glutamate (PBLG).

phrase 'frustrated chain packing'.) One approach is to introduce a kink into the rod by using naphthalene moieties in the backbone. Another approach is to introduce non-para linkages so that the molecule rod is less straight. Two other approaches are to add bulky substituents to the aromatic rings (another example of frustrated chain packing), and to incorporate flexible linkages into the backbone.

The rheology of liquid crystal polymers was reviewed by Wissbrun.[8] A more recent review is from 1996,[10] particularly Chapter 4 by Cogswell and Wissbrunn, which makes the point that LCP 'changes the rules which govern the balance between ease of processing and service performance' compared to conventional thermoplastics reinforced by the addition of discrete fiber fillers. These reviews describe the essential features of the usual response of thermotropic polymers to shear, namely shear-thinning,

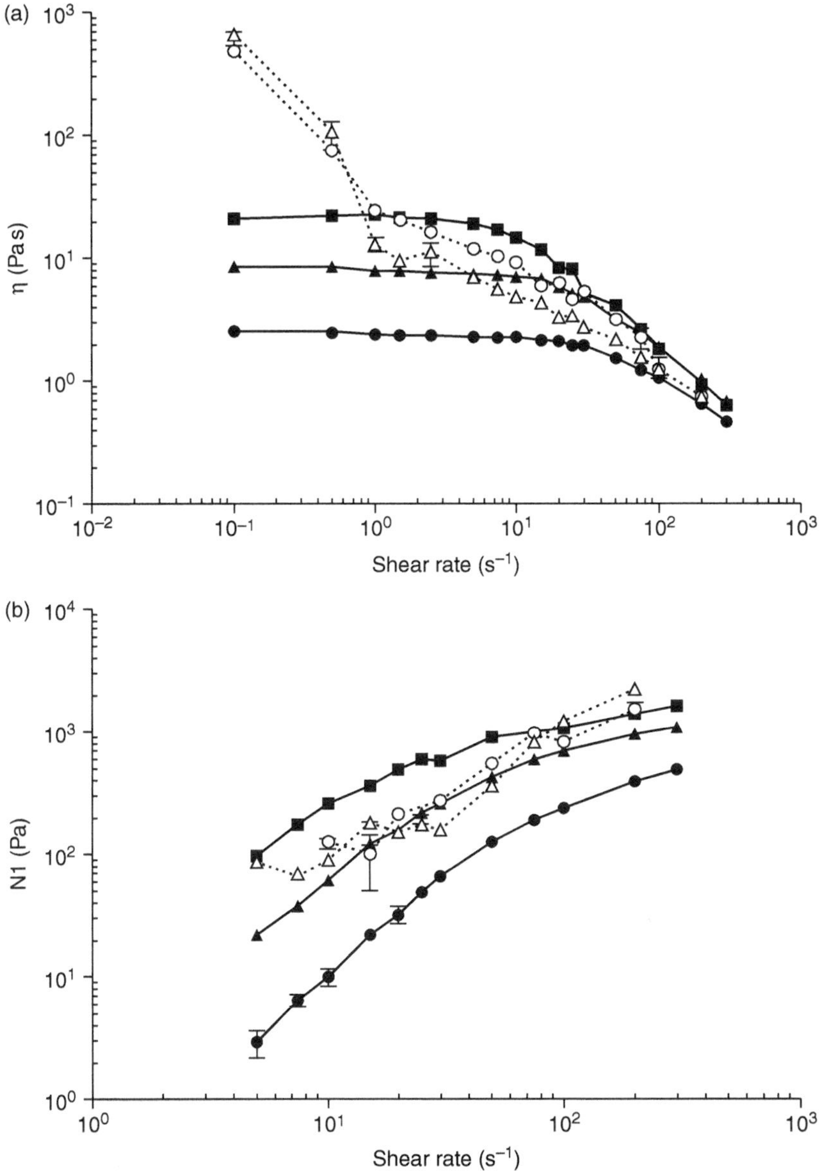

5.4 Normal stress behavior for PBLG in m-cresol: (a) viscosity, (b) normal stress. Different symbols represent data taken at different temperatures.

lower viscosity than isotropic polymers, and conventional (negative) thermal coefficient of viscosity. However, thermotropic polymers also reveal some rheological surprises. There are no examples of steady-state negative normal stress, but there are observations of transient negative normal

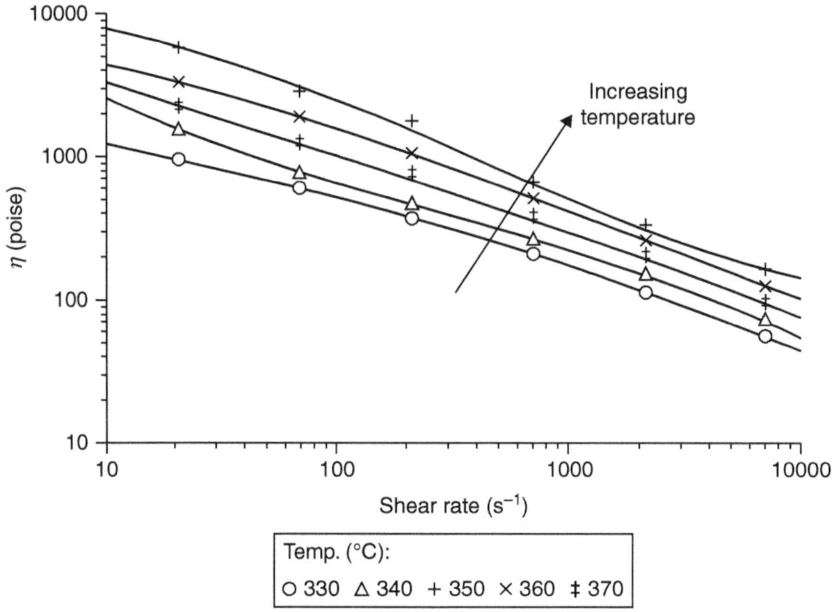

5.5 Flow curves at various temperatures for HIQ 35.

stress.[8] There is also a report of unconventional (positive) thermal coefficient. Viscosity was observed to increase with temperature over a broad range of temperatures and shear rates for a copolymer of hydroxybenzoic acid (HBA), isophthalic acid (IA), and hydroquinone (HQ) called HIQ[29] as seen in Fig. 5.5 for HIQ 35. This unusual phenomenon was confirmed[28] and has application in LCP blends, as we will see later in Section 5.6.2.

5.5 Blends of isotropic and anisotropic thermotropic polymers

The concept of blending of isotropic polymers with LCP's goes even further in the direction of 'changing the rules', as quoted in Section 5.4.2, as a strategy for engineering materials which strike exactly the right balance of properties and processability. Since the major component (the matrix) is the isotropic polymer, the overall blend exhibits many of its properties (not the least of which is cost) or at least is heavily biased in that direction. Furthermore, the degree of freedom traversed in optimizing the material is the concentration of the LCP phase, which is easily controlled in compounding or even at the last minute by dry-blending in the hopper, which is certainly easier than varying chemistry or molecular weight. An indispensable reference which reviewed the state of the art in 1990 is Dutta, Fruitwala,

Kohli, and Weiss.[7] This paper also considered small-molecule liquid crystals as additives to conventional thermoplastics.

5.5.1 Miscibility and compatibility

When blending two or more polymers, the question of miscibility comes up. And if the two components are immiscible, then the next questions to be asked are those of adhesion and compatibilization. Polymers in most cases are not miscible because of low enthalpy of mixing, but examples of miscible blends containing an isotropic and a thermotropic polymer are known. One recent example is a study elucidating the phase diagram of polycarbonate with HBA-HNA copolyester[30] which used optical microscopy and transmitted light intensity analysis to determine two separate phase separation temperatures at which 'slight' and 'pronounced' phase separation occurred. They also used NMR and FTIR to establish that no transesterification was occurring. Their Figure 13 shows a phase diagram with temperature from 220°C to 300°C and LCP content between 20 wt % and 60 wt % exhibiting three well-defined regions.

Blends of polycarbonate with the HIQ polymer described in Section 5.4.2 were found to be miscible.[30] The authors observed by thermal and thermomechanical analysis that the T_g of the blend was greater than that of either pure component, and indeed the T_g of the HIQ component was absent in a blend containing as much as 75% HIQ. They consider the combination of miscibility of LCP with a thermoplastic and a non-intermediate T_g to constitute a novel polymer blend pair.

The minimum performance which is hoped for in a polymer blend is to follow the Rule of Mixtures, which is simply a volume averaged result based on the composition. If both components are polymers, then volume average and weight average are about the same. In many cases the performance of the blend falls below the Rule of Mixtures, due to poor adhesion, degradation or other chemistry during processing, details of orientation and morphology, and a host of other unsurprising problems which can lead to disappointing performance. The surprising and hoped-for result is performance which exceeds that predicted by the Rule of Mixtures, which would represent a synergy. Better yet is performance which exceeds that of either pure component.

5.5.2 Mechanical properties of LCP/polymer blends

The motivation behind much of the work on LCP blends is the desire to achieve improved mechanical properties in a useful regime of cost, performance, and processability. The early work in this area took place shortly after the advent of thermotropic polymers and was an attempt to capitalize on the desirable points of thermotropic polymers while avoiding some of the drawbacks (cost, poor transverse properties, weld lines, etc.)

Synergies in mechanical properties were reported by Isayev and Modic.[31] Dynamic mechanical measurements showed local maxima above the Rule of Mixtures in storage modulus for two LCPs blended with polycarbonate. Young's modulus of extruded strands showed local maxima above the Rule of Mixtures, especially for 'unsupported' strands, leading to the possibility that this was a drawing effect. However, local maxima in Young's modulus were also observed for compression molded films and injection-molded bars. The most striking observation was that of transverse strength in injection-molded bars being higher for blends than for either base polymer (p. 166 and Figure 11 of reference 31). Synergy in viscosity behavior would be the observation of viscosity lower than that predicted by the Rule of Mixtures, or even lower than either neat polymer.

5.6 Processability of LCP/thermoplastic blends

As referred to earlier, LCP/thermoplastic polymer blends represent a route to novel materials that 'change the rules' when it comes to the balance and tradeoffs between processability and properties. In addition to the ability of these low-viscosity blends to fill long flow paths (giving greater flexibility in gate location) and thin-wall sections (conserving material) there are other considerations such as energy consumption during processing, abrasion of tools, and surface quality. Abrasion of screws, runners, and molds is particularly important in the case of medical devices, which is an application area that could tolerate the relatively high cost. Metal contamination is clearly unacceptable for medical devices, and surface quality is also critical.

5.6.1 Viscosity ratio

A thorough study of the effect of viscosity ratio (LCP to TP) was carried out by Fekete *et al.*[32] Although some authors have asserted that a viscosity ratio of less than one (LCP less viscous than TP at the processing condition) would be optimal, this was not borne out in this study. Of course 'processing conditions' is a broad term, which includes temperature, pressure, shear rate and possibly elongation rate as well. This paper compared flow-direction and transverse strength for a variety of TPs and LCP contents, and concluded with this rule of thumb: 'The results of the experiments revealed that the reinforcing effect is directly related to the viscosity decrease of the blends compared to the neat thermoplastics.'

A particularly interesting test of the above rule of thumb would be blends containing the HIQ polymer described in Section 5.4.2 in which the unusual positive thermal coefficient of viscosity was attributed to the coexistence of isotropic and anisotropic domains of the coploymer, as determined by the distribution of copolymer ratios and therefore local chain stiffness. As temperature increased, the fraction of the high viscosity isotropic phase increases at the expense of the low-viscosity anisotropic phase. This polymer by itself as the LCP component would open a degree of freedom in varying the viscosity ratio. Furthermore, blending this HIQ LCP with more conventional LCPs with which it is miscible[33] would expand the window of control of the LCP component.

5.6.2 Spiral flow

The spiral flow test is commonly used to pragmatically determine the ability of a polymer to tolerate long flow lengths and form thin-walled injection-molded parts. We have mentioned that a particular LCP called HIQ exhibits the unusual property of positive temperature dependence of viscosity, the opposite of all known isotropic thermoplastics. This phenomenon was exploited in LCP/thermoplastic polymer blends to provide self-stabilized viscosity and greater resistance to process drift.[34] Figure 5.6 shows viscosity independent of temperature for a 35 wt% HIQ/Ultem blend, and the stabilizing effect on spiral flow length when the mold temperature was abruptly increased by 30°C. This caused a 75% increase in spiral flow length for unblended Ultem after 20 shots but only a 20% increase in length for the blend.

5.6.3 Weld lines

A well-known characteristic of chopped-fiber composites and LCP is that they are extremely sensitive to weld lines in injection molding, so that care must be taken to move the weld lines to non-critical areas within the part. A study of butt-type weld lines compared LCP/TP blends with pure matrix polymer for a number of TPs and determined that there was significant reduction in strength relative to the non-weldline case, but that there were blends in which the weldline strength was nevertheless greater than the pure TP. In every case, the tensile modulus, which is often the more important property for engineering thermoplastics, was greater for the LCP/TP blend weldline than for the maximum observed for the pure TP (in some cases the

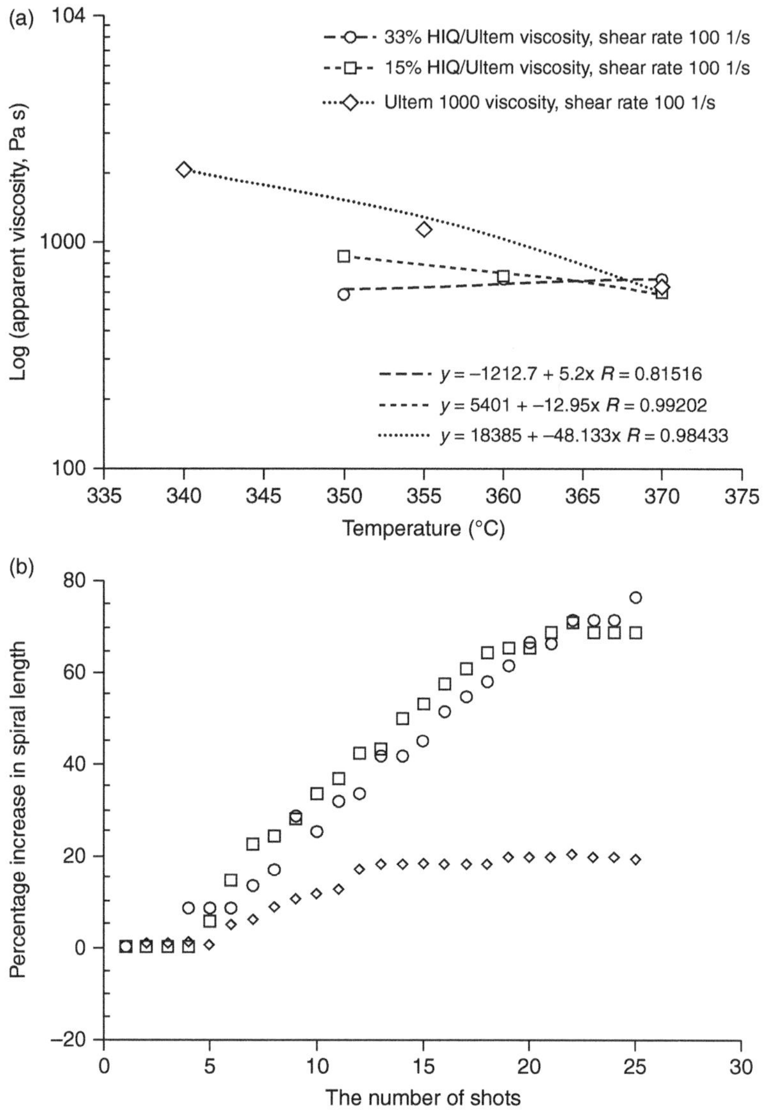

5.6 Pure Ultem compared to HIQ/Ultem blends. (a) Temperature dependence of viscosity. (b) Spiral flow length following step increase in mold temperature.

pure TP had higher modulus for the weldline case than for the non-weldline case). This study did not compare with filled grades, which of course would be the more commercially significant comparison. Weld lines for blends of immiscible polymers in which neither component is an LCP were found to be lower than either pure component.[35]

5.7 Structure–property relationships of LCP blended materials

Polymers in general have properties which depend strongly on orientation of the chains. LCPs in particular are extremely sensitive to processing conditions which give rise to morphologies that may or may not be favorable for developing favorable mechanical properties. If the LCP is present in spherical or low-aspect ratio domains, then it is used inefficiently for reinforcement, which is particularly problematic given their high cost. Significant effort and sophistication are required to achieve a worthwhile improvement in properties by managing the flow during processing, and the adhesion of the reinforcing domains to the matrix. Another factor to consider is that the highly oriented LCP component can act as a nucleating species to encourage the flexible polymer matrix to crystallize in an oriented rather than a random fashion. The notion is similar to the concept of 'row-nucleation', which was studied to such spectacular effect by Keller.

A distinction must be made between extruded and fiber-spun articles, which benefit from uniform shear and elongation flow, both of which are highly efficient at orienting LCP domains, and the complex flows experienced in injection-molded articles. A recent paper on the structure-property relationships of PEN (Polyethylene Naphthalate)/LCP blend fibers used a novel CO_2 laser heating method during fiber draw to heat the fiber uniformly rather than outside-in, and achieved synergistic effects.[36] The paper included SEM micrographs of well-oriented and well-adhering LCP fibrils, and an example of LCP fibrils is shown in Fig. 5.7. Interestingly, naphthalene dicarboxylate (NDC), a key monomer for PEN, has also been proposed as a lower-cost monomer for LCPs.

Injection-molded articles tend to exhibit skin-core and multi-layer morphologies in which the LCP domain aspect ratio varies considerably with location within the part. The hierarchy of structures found within LCP/TP blends was elucidated by Silverstein *et al.*,[37] who said 'the microdomains form domains, the domains form microlayers, the microlayers form sublayers, and the sublayers form a layered structure'. The structural hierarchy in PET/CLP blends was characterized by Narh[5] who found good agreement of fibril length observed in sublayers with a numerical prediction.

The transition between low-aspect ratio and high-aspect ratio domains can be quite abrupt, as shown in reference 13. In those portions of the part where the flow is favorable to the formation of high-aspect ratio domains, the 'forest' of LCP 'needles' emerging from the surface can be quite spectacular, especially considering that the domains did not exist prior to the formation of the part, but were formed *in situ*.

Of course the adhesion between the LCP domains and the matrix must be poor for the elongated domains to pull out and appear needle-like, rather

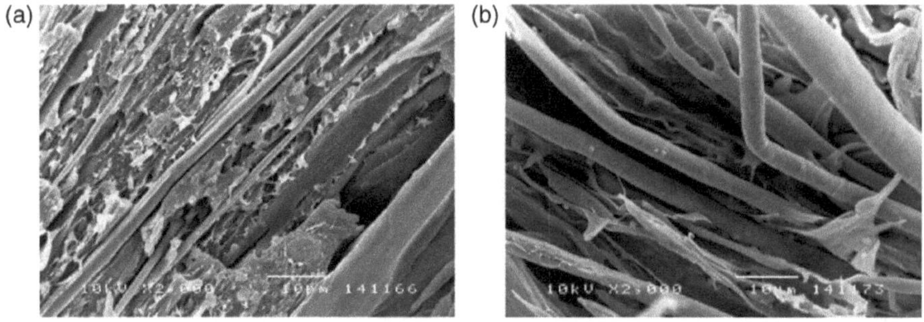

5.7 SEM micrographs of LCP fibrils, both skin (a) and core (b) regions.

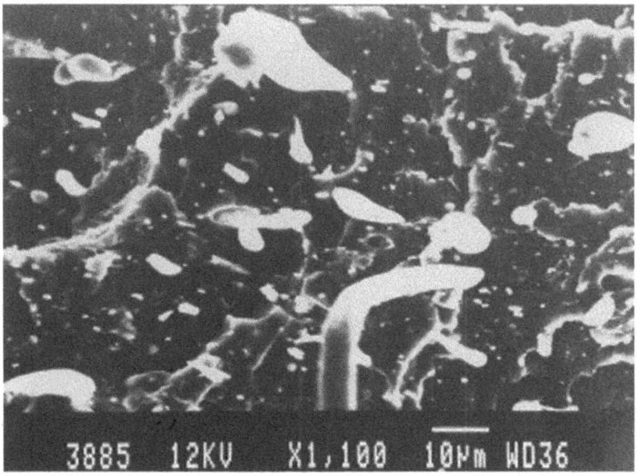

5.8 Example of good adhesion in LCP blend.

than breaking, so this vivid image is not necessarily a favorable one. A better image from the point of view of useful reinforcement is one of broken, but clearly elongated, domains, in which the flow-induced internal structure of the LCP domain may be visible in a brittle fracture surface of the domain in the blend product.

In other cases the adhesion between matrix and LCP reinforcement is good. The next SEM micrograph, Fig. 5.8, shows a rather similar forest of tapered needles, but in this case the tapering was due to drawdown during break (as opposed to brittle fracture) and not to pull out of poorly adhering domains, since the adhesion is seen to be excellent.

Recent work in *in situ* composites has included the incorporation of single and multi-wall carbon nanotubes (SWCNT and MWCNT). Notably, a low-cost alternative exists in the form of thermal chemical vapor deposition-grown carbon nanofibers (CNF) which are slightly inferior due to greater diameter and shorter length, but are available in kg quantities.[38] Carbon nanotubes and LCP seem to be highly complementary because both have a high aspect ratio. LCP is thought to promote good dispersion of CNF, which is necessary to develop good mechanical properties efficiently. In addition, the good dispersion promotes the formation of a conductive network by achieving a percolation threshold at lower CNF loading, leading to improved electrical conductivity at lower cost.

The adhesion between CNF and LCP appears to be favorable, or can be made so using surface treatment of the CNF,[39] leading to the observation of 'bridging' nanofibers which transfer stress from the matrix to the LCP reinforcement, particularly under conditions where the LCP domains are also undergoing fibrillation.[40]

5.8 Commercial LCP blends

This section contains a description of commercial activity in LCP blends. In addition, a discussion of patent activity on the topic is provided.

5.8.1 Patents

Much of the early publication in this field was in the form of patents. There was a flurry of patent activity in the early 1980s in the area of LCP/thermoplastic polymer blends, starting with Takayanagi at Asahi Chemical.[41,42] Early contributions were also made by Cogswell, Griffin and Rose at ICI.[43,44] A number of patents came out of Celanese, including a string by Froix combining LCP with a number of different matrix polymers.[45–52] Most of these were relatively straightforward composition of matter patents, though reference 52 was unusual in that in addition to the composition of the blend it specified the size and aspect ratio of the LCP domains within the blend. Isayev at University of Akron also produced at least one patent in the area at this time.[53]

Celanese remained active in this area and produced some interesting technology in the mid-1990s. Many of these involved blends in which both components were anisotropic (Section 5.4), but Makhija *et al.*[54] describe a technique for LCP/isotropic thermoplastic blends in which components are fed sequentially into a mixing die to give additional control over the morphology. Either the LCP or the isotropic polymer can be first, followed by the other component. In some cases solid polymer is fed into an extruded stream of melt. The patent gives numerous examples of PET first vs. LCP first vs. simultaneous feed, and shows improved mechanical properties.

A relatively recent patent from Foster-Miller[55] describes a multiaxially oriented film in which the LCP component of the blend provides barrier properties and controls CTE, while allowing operations such as heat-sealing and microwave heating of food. This was the technology licensed by the Superex spin-off discussed below.

A quite recent Celanese patent is that of Jester,[56] which describes blending a 'stretchable' LCP into a thermoplastic so that the reinforcement can deform with the matrix and thereby maintain adhesion during post-processing steps. One of the most recent patents we found in the area of LCP blends is actually an LCP/LCP blend rather than an *in situ* composite.[57] It appears that this blend was developed for a specific application (housing for air-conditioner with high-speed fan) which required rigidity, long flow paths, thin walls, and low processing temperature for excellent 'weldability'. It demonstrates the progress which continues to be made in filling specific niches by engineering materials via blending rather than chemistry. It would be an interesting exercise to determine if this specific housing could have been made equally well but at lower cost with an *in situ* composite blend.

5.8.2 Commercial products

LCP blends have 'shown promise' for many years, and yet there have been relatively few actual commercial applications. A great disadvantage of LCPs is their high cost relative to glass fiber reinforcement, which makes it hard to compete purely on mechanical properties. It therefore becomes necessary to compete with conventional plastics on the basis of other properties which confer high value. Examples are electrical properties, gas barrier properties, and controlled thermal expansion coefficient. Pure LCPs excel in these areas, and blends containing a significant amount of LCP can improve conventional thermoplastics significantly.[58] The improvement in processability is a bonus which may tip the balance in favor of an LCP-modified grade, or in some cases make an application possible which otherwise might not be, such as thin-walled objects. CTE control is a major issue with fiberoptic cables, and approaches using LCP have been patented. A blend of LCP with Nylon 12 was reported to show rapid decrease in CTE with LCP content, reduced to 25% of the pure Nylon value as LCP content reached 20%, near-zero CTE at LCP content of 40% and slightly negative CTE with even greater LCP content.

Some descriptions from the trade literature of the 1990s indicated that commercial products were being launched. Although some of the blends described were development projects, a few sound as if they were commercial at the time. Hoechst-Celanese (actually its Polyplastics joint venture with Daicel in Japan) had two Vectra grades X037 and X039 which were PC/LCP blends, neither of which have survived.

A spin-off from Foster-Miller in Waltham MA called Superex Polymer marketed a material in the mid-1990s for \$2–\$3/lb which was described as a PET/LCP blend called 'Super PET' in a Nasa Tech Report link.[59] See also reference 55. This material was intended to prevent gas ingress into food, egress from carbonated beverages (especially beer), and water penetration into power cables.[60] This material was offered as a licensable technology patent until at least 2000 and possibly beyond. The company was folded back into Foster-Miller in 2000, and Foster-Miller itself was acquired by QinetiQ in 2004.[61] Hoechst-Celanese has become Ticona, and they have two grades of PolyPhenylenSulfide (PPS)/LCP, sometimes described as a blend and sometimes as an alloy.[62] It is not clear from the product literature, but the matrix component is indeed the PPS.[63]

5.8.3 LCP/LCP blends

This section of the chapter deals with blends in which both components are LCP. Among the series of patents awarded to Froix in the 1980s was US 4,267,289, issued in May 1981.[64] This is a composition of matter patent which does not indicate the benefit but simply claims blends in which one LCP component uses a naphthalene moiety to render the aromatic polyester tractable, whereas the other component uses methyl, chloro, or bromo groups on the aromatic ring.

Another early work in this area is that of DeMeuse and Jaffe from Celanese[65] who used several techniques such as rheology, X-ray diffraction, and calorimetry to demonstrate that a model system consisting of blends of two LCP polyesters of differing HBA/HNA ratios were in fact not miscible. This is in contrast to low molecular weight liquid crystals in which two liquid crystals which form the same type are expected to be miscible. This paper also demonstrated that transesterification did not occur at an appreciable rate under the conditions used for melt rheology measurements. They also discuss the concept that for copolymers in which the distribution of copolymer ratio is extremely wide, there could be phase separation within a nominally homogeneous copolymer. This is precisely the mechanism invoked to explain the anomalous temperature dependence of viscosity of HIQ LCP.[29]

Another study of miscibility of LCP/LCP blends also used the HIQ LCP, blended with Vectra A950 thermotropic polyester.[30] This paper confirmed the biphasic nature of HIQ and that HIQ is miscible with Vectra. It determined that the Cox-Merz rule (equivalence of dynamic and steady shear viscosity) was obeyed for the HIQ 45 and blends with Vectra A950 and the anomalous temperature of viscosity dependence was manifested for pure HIQ. They did not, however, indicate what blend composition of HIQ/Vectra manifested viscosity independent of temperature (since Vectra has conventional temperature dependence of viscosity). The same paper also determined that HIQ is miscible with polycarbonate, but the Cox-Merz rule is not followed. They concluded that in both systems the overall behavior is dominated by the less ordered component, whether it be the PC component or the isotropic component of biphasic HIQ.

US Patent 5,393,848, Charbonneau *et al.*, is a Celanese patent from 1995 which discusses *in situ* polymerization of HIQ to form an LCP/LCP blend *in situ*.[66] The idea here is to create HIQ by polymerizing the monomers within a matrix of molten LCP, for example Vectra. This is a clever way to gain better control over the polymerization process and avoid formation of the problematic biphasic melt. The attention paid to HIQ is because it is produced from three relatively common and low-cost monomers, hydroxybenzoic acid, isophthalic acid, and hydroquinone, avoiding the use of expensive specialty monomers. This method is also proposed as a method to create LCPs which would not be tractable if synthesized conventionally.

A very recent LCP/LCP blend patent[57] was already mentioned in Section 5.5.1. A companion patent also issued to Ueno Fine Chemicals is US 7,842,760, issued 20 November 2010, claims essentially the same blend but with emphasis on its dielectric properties.[67]

5.9 Conclusions and future trends

The rule of thumb ('... the reinforcing effect is directly related to the viscosity decrease of the blends compared to the neat thermoplastics') stated by Fekete *et al.*[32] is intriguing but of course too simple and limited. There is a need for a more general theory or predictive model of LCP blends which would allow input of rheological parameters, parameters related to adhesion, orientability, degradation temperature, etc. and of course the cost of the components, and allow one to compare the expected material to commercially available grades of glass-filled thermoplastics, etc.

We feel confident that given the dual benefit of reinforcement and processability improvement, there are high value niches which would be best filled by an LCP blend, particularly a relatively low-cost variant. The potential low cost and unusual rheology of HIQ polymers opens up interesting scientific and commercial possibilities. It would give an opportunity to control viscosity ratio in a novel way. It could be used either alone, or blended with Vectra, since the two seem to be miscible. We would like to see a resurgence of interest in LCP blends and the introduction of commercially successful *in situ* composite products after all these years.

5.10 References

1. Janssen, L.P.B.M., Boetes, R., Smith, J.M., 'Self Reinforcing Blends', Polymer Extrusion II Confer, *Temperature Dependence of Viscosity of Thermotropic Polyesters*, G. Kiss, J Rheo (eds) V30, pp. 585–599 (1986).

2. Collyer, A.A., 'The morphology and rheology of liquid crystal polymer blends', Chap 6 *Rheology and Processing of Liquid Crystal Polymers* (ISBN: 9780412596407) D.A. Acierno, A.A. Collyer (eds), pp. 86–134, Books Express, (Saffron Walden, ESS, United Kingdom) (1996).
3. Kiss, G., Kovacs, A.J., Wittmann, J.C., 'In-situ polymerization in binary polymer/monomer mixtures: a novel route to polymeric composites', *Journal of Applied Polymer Science* **V26** 2665 (1981).
4. *http://thomsonreuters.com/products_services/science/science_products/a-z/web_of_science/*.
5. Narh, K.A., 'Liquid crystalline polymer blends as a route to self-reinforcing composites', *J. Reinforced Plastics and Composites* **V28**, 1957–1973 (2009).
6. Baird, D.G., Blizard, K., 'Non-Isothermal crystallization kinetics and thermal stability of the in situ reinforced composite films based on thermotropic liquid crystalline polymer and polypropylene', *J. Therm Anal Calorim* **V103**, 1017–1026 (2011).
7. Dutta, D., Fruitwala, H., Kohli, A., Weiss, R.A., 'Polymer blends containing liquid crystals', *Poly Eng Sci* **V30**, 1005–1018 (1990).
8. Wissbrun, K.F., 'Rheology of liquid crystal polymers: a review', *J. Rheology*, **V25**, 619 (1981).
9. Acierno, D.A., Collyer, A.A., *Rheology and Processing of Liquid Crystal Polymers* (ISBN: 9780412596407), Books Express, (Saffron Walden, ESS, United Kingdom) (1996).
10. Cogswell, F.N., Wissbrun, K.F., 'Rheology and processing of liquid crystal polymer melts', Chap 4 *Rheology and Processing of Liquid Crystal Polymers* (ISBN: 9780412596407) D.A. Acierno; A.A. Collyer eds., pp. 86–134, Books Express (Saffron Walden, ESS, United Kingdom) (1996).
11. Chung, T.-S., 'How well do various blends of LCP and nylon 12 work?', *Plastics Engineering* p. 39 (1987).
12. Blizard, K., Baird, D.G., 'Morphology and rheology of polymer blends containing a LCP copolyester', *Poly Eng Sci* **V27**, 653–657 (1987).
13. Kiss, G., 'In situ composites: blends of isotropic polymers and thermotropic liquid crystalline polymers', *Poly. Science* (RAPRA) London, 5/12–13/1982, paper 25.1 (1982).
14. Weiss, R.A., Wansoo, H., Nicolais, L., 'Novel reinforced polymers based on blends of polystyrene and a thermotropic liquid crystalline polymer', *Poly Eng Sci* **V27**, 684–691 (1987).
15. Kiss, G., 'Topics in liquid crystal rheology', Malvern Webinar, 51 slides, 36 minutes, *http://www.malvern.com/malvern/ondemand.nsf/webcasts?opennavigator'year=2010* or *https://www.brainshark.com/malvern/vu?pi=zEGzZg27lzB28z0*.
16. Gray, G.W., Harrison, K.J., Nash, J.A., 'New family of nematic liquid crystals for displays', *Electronics Lett.* **V9**, 130 (1973).
17. de Gennes, P.G., Prost, J., *The Physics of Liquid Crystals*, ISBN 0-19-852024-7, Clarendon Press, (Oxford) (1993).
18. Porter, R.S., Johnson, J.F., *Rheology*, 4, F.R. Eirich, Editor, Academic Press (New York) (1979).
19. Bawden, F.C., Pirie, N.W., Bernal, J.D., Fankuchen, I., 'Liquid crystalline substances from virus-infected plants', *Nature* **V138**, 1051–1052 (1936).

20. Lauffer, M.A., 'Optical properties of solutions of tobacco mosaic virus protein', *J. Phys. Chem.* **V42**, 935–944 (1938).
21. Griffin, A., Cox, B., *Brit Poly J* **V12**, 147 (1980).
22. Kwolek, S.L., Morgan, W.P., Sorenson, R.W., 'Process of Making Wholly Aromatic Polyamides', DuPont, US Patent 3063966, 11/13/1962.
23. Kwolek, S., 'Optically Anisotropic Aromatic Polyamide Spinning Dopes', DuPont, US Patent 3,671,542, June 20, 1972.
24. Miller, W.G., 'Stiff chain polymer lyotropic liquid crystals', *Ann Rev Phys Chem* (1978) *http://www.annualreviews.org/doi/abs/10.1146/annurev.pc.29.100178.002511*.
25. Flory, P.J., *Proc. Royal Society of London*, **A234**, 60 (1956).
26. Hermans, J., *J. Colloid Sci* **V17**, 638 (1962).
27. Iizuka, E., *Mol Crys Liq Crys*, **25**, 287 (1974).
28. Kiss, G., 'Flow-Induced Phenomena of Lyotropic PLCs' in *Mechanical and thermophysical properties of polymer liquid crystals*, W. Brostow Editor, Chapman and Hall, (London) (1998).
29. Kiss, G., *Polym. Eng Sci* **V27**, 410–423 (1987).
30. Lee, K.-W.D., Chan, P.K., Kamal, M.R., 'Determination of the phase diagram of HBA-HNA liquid crystalline polymer/polycarbonate blends', *J. Appl. Poly. Sci.* **V111**, 396–407 (2009).
31. Isayev, A.I., 'Self-reinforced melt processable polymer composites: extrusion, compression and injection molding', *Polymer Composites* **V8**, 158–175 (1987).
32. Fekete, E., Foldes, E., Lengyel, Z., 'Effect of viscosity ratio of the components of thermoplastic and liquid-crystalline polymer blends on their properties', *Composite Interfaces* **V13**, 737–755 (2006).
33. Hsieh, T.-T., Tiu, C., Simon, G.P., Wu, R.Y., 'Rheology and miscibility of thermotropic liquid crystalline polymer blends', *J. Non-Newtonian Fluid Mech.* **V86**, 15–35 (1999).
34. Wei, K.H., Kiss, G., 'Liquid crystalline polymer blends with stabilized viscosity', *Poly Eng Sci* **V36**, 713–720 (1996).
35. Kelley, R.A. ley, Barlow, J.W., Paul, D.R., *Poly Eng Sci* **V20**, 364–372 (1980).
36. Kim, S.Y., Kim, S.H., Lee, S.H., Youn, J.R., 'Internal structure and physical properties of thermotropic liquid crystal polymer/poly(ethylene 2,6-naphthalate) composite fibers', *Composites Part A* **V40**, 607–612 (2009).
37. Silverstein, M.S., Hiltner, A., Baer, E., 'Hierarchical structure in PET/LCP blends', *J Appl Poly Sci.* **V43**, 157–173 (1991).
38. Kumar, S., Rath, T., Mahaling, R.N., Das, C.K., 'Processing and characterization of carbon nanofiber/syndiotactic polystyrene composites in the absence and presence of liquid crystalline polymer', *Composites A* **V38**, 1304–1317 (2007).
39. Nayak, G.C., Rajasekar, R., Das, C.K., 'Effect of SiC coated MWCNTs on the thermal and mechanical properties of PEI/LCP blend', *Composites A* **V41**, 1662–1667 (2010).
40. Nayak, G.C., Rajasekar, R., Das, C.K., 'Effect of modified MWCNT on the properties of PPO/LCP blend', *J. Mater. Sci.* **V46**, 2050–2057 (2011).
41. Takayanaki, M., Kajiyama, T., 'Polymer composite material', Asahi Chemical, UK Pat Appl 2,008,598 6/6/79.
42. Takayanagi, M., Kajiyama, T., 'Polymer composite material', Asahi Chemical, US Patent 4,228,218 10/14/80.

43. Cogswell, F.N., Griffin, B.P., Rose, J.B., 'Compositions of melt-processable polymers having improved processability, and method of processing', Imperial Chemicals, Eur. Pat. Appl. EP0030417 (A1) 6/17/81.
44. Cogswell, F.N., Griffin, B.P., Rose, J.B., 'Compositions of melt-processable polymers having improved processability', Imperial Chemicals, US Patent 4,386,174, 5/31/81.
45. Cincotta, D., Berardinelli, F., 'Blend of polyalkylene terephthalate and wholly aromatic polyester and process for producing the same', Celanese, US Patent 4408022, 10/4/83.
46. Cincotta, D., Berardinelli, F., 'Process for forming blend of polyalkylene terephthalate and wholly aromatic polyester which exhibits an anisotropic melt phase', Celanese, US Patent 4451611, 5/29/84.
47. Froix, M.F., 'Blend of polycarbonate and wholly aromatic polyester', Celanese, US Patent 4,460,735, 7/17/84.
48. Froix, M.F., 'Blend of polyarylene sulfide and wholly aromatic polyester', Celanese,US Patent 4,276,397, 6/30/81.
49. Froix, M.F., 'Blend of sulfone polymer and wholly aromatic polyester', Celanese, US Patent.
50. Froix, M.F., 'Blend of polyalkylene terephthalate and wholly aromatic polyester', Celanese, US Patent 4,489,190, 12/18/84.
51. Kiss, G., 'Blends of an aromatic polyesteramide exhibiting an anisotropic melt phase', Celanese, US Patent 4,567,227 1/28/86.
52. Kiss, G., 'Melt blend of non-thermotropic and thermotropic wholly aromatic polyesters', Celanese, European Patent Application EP 0169947(A1) 2/5/86.
53. Isayev, A., Modic, M.J., 'Liquid Crystal Fiber-Reinforced Polymer Composite and Process for Preparing Same', University of Akron, US Patent 4,728,698, 3/1/88.
54. Makhija, S., Lundberg, G.L., Stamatoff, J.B., 'Method for Making Blends of Liquid Crystalline and Isotropic Polymers', Celanese, US Patent 5,541,240, 7/30/96.
55. Rubin, L.S., Blizard, K.G., Haghighat, R.R., Lusignea, R.W., 'Extruded Thermoplastic, Liquid Crystalline, Polymers and Blends Thereof Having a Planar Morphology', Foster-Miller, Inc., US Patent 5,981,007, 11/9/1999.
56. Jester, R.D., 'Blends of Stretchable Liquid Crystal Polymers with Thermoplastics', Ticona, US Patent 6,660,182 (12/9/2003).
57. Terada, H., Asahara, M., Kato, H., 'Liquid-Crystalline Polymer Blend and Composition Comprising the Same', Ueno Fine Chemicals, US Patent 7,531,222 (5/12/2009).
58. Roggero, A.D., 'Applications of Thermotropic Liquid Crystal Polymers and Thermotropic Liquid Crystal Polymer Blends' in *Thermotropic Liquid Crystal Polymer Blends*, F.P. La Mantia, ed., Technomic, (Lancaster, PA) (1993) pp 157–175.
59. 'Advanced Polymers for Practical Use', NASA Tech Report Document ID: 20020063434 *http://ntrs.nasa.gov/archive/nasa/casi.ntrs.nasa.gov/20020063434_2002103549.pdf*.
60. Lusignea, R., 'LCPs are Moving into Blow Molded Bottles, Look for LCPs in food trays, fuel tanks, power cables', *Plastics Technology*, p. 11 (June 1996).
61. Richard Lusignea, QinetiQ USA, personal communication, 9/1/11.

62. 'Vectra Product Line', Ticona GmBH, *http://www.ticona.com/vectrabr_e.pdf* (October 2007).
63. James Shepherd, Ticona GmBH, Personal Communication 8/31/11.
64. Froix, M.F., 'Blend of wholly aromatic polyesters', US Patent 4,267,289 5/12/81.
65. DeMeuse, M.T., Jaffe, M., 'Model system for liquid crystal polymer blends', *Mol Crys Liq Crys* **V157**, 535–566 (1988).
66. Charbonneau, L.F., Gupta, B., Sawyer, L.C., Linstid, H.C., Shepherd, J.P., 'Process for forming improved liquid crystalline polymer blends', Celanese, US Patent 5,393,848 2/28/95.
67. Yonezawa, S., Asahara, M., Fujiwara, H., Terada, H., Kato, H., 'Liquid-crystalline polyester blend', Ueno Fine Chemicals, US Patent 7,842,760. 11/30.

6
Polysulfones as a reinforcement in high temperature polymer blends

M. T. DeMEUSE, MTD Polymer Consulting, USA

DOI: 10.1533/9780857099013.165

Abstract: Polysulfones are a class of high temperature polymers that have applications in emerging fields such as membranes. The high temperature nature of the polysulfones allows them to be used in demanding applications that other polymeric materials cannot satisfy. However, as new applications continue to emerge, the properties of the polysulfone need to be modified appropriately. One approach to achieving that goal is through blending with other polymers, both thermoplastics and thermosets. This chapter provides a summary of those blending efforts and suggests additional areas for further research and development work. The use of polysulfones in both membrane applications and as impact-modifiers for thermosetting materials is discussed.

Key words: polysulfone, membrane, thermoset, impact-modified, amorphous, separation.

6.1 Introduction

The general term 'polysulfone' describes an entire family of thermoplastic polymers which are melt processable. The polymers are particularly known for their toughness and stability at high temperatures. They contain the subunit aryl-SO_2-aryl, the defining feature of which is the sulfone group. This chapter will discuss blends of polysulfone polymers with other polymers, both thermoplastics and thermosets, and the results obtained from the production of such mixtures.

6.2 Structure and properties of polysulfone

The general chemical structure of polysulfone is shown in Fig. 6.1. It is an amorphous polymer that has a glass transition temperature or T_g of about 185°C. Due to the aromatic backbones present in the chemical structure, polysulfones typically possess high stiffness and strength. Articles made from polysulfone have excellent transparency features.

Polysulfone is highly resistant to mineral acids, alkali and electrolyte materials. It is resistant to oxidizing agents, surfactants and hydrocarbon

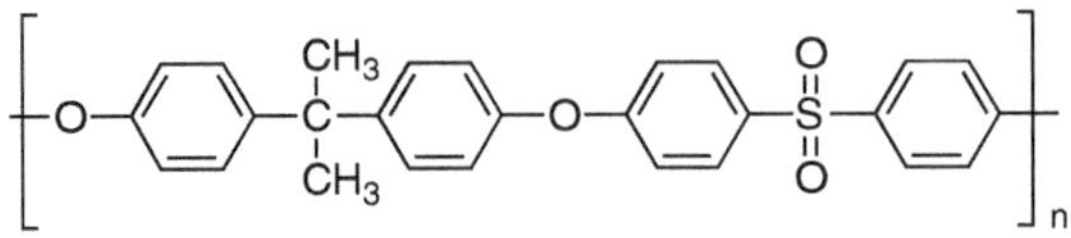

6.1 Structure of polysulfone polymer.

oils. It is not resistant to low polarity organic solvents such as ketones, chlorinated hydrocarbons and aromatic hydrocarbons. It is also stable in aqueous acids and bases and many non-polar solvents. However, it is soluble in both dichloromethane and methylpyrrolidone.

Polysulfone has one of the highest service temperatures of all melt-processable thermoplastics. Its resistance to high temperatures gives it a role as a flame retardant material, without compromising the strength in the way that usually results from the addition of flame retardants. Its very high hydrolytic stability allows for its use in numerous medical applications that require autoclave and steam sterilization. However, it also has low resistance to some solvents, as already noted, and does undergo weathering effects.

The use of polysulfone allows for the easy manufacturing of membranes with controllable size of pores down to 40 nanometers. Such membranes are used in applications such as hemodialysis, waste water recovery, food and beverage processing, and gas separation. Also, polysulfone can be reinforced with glass fibers to form composite materials. Further, it is used as a dielectric material in capacitor applications.

One of the primary rationales for producing blends of polysulfone with other polymers is to use the polysulfone to impart separation and membrane capabilities to the material and for the second polymer to provide higher temperature performance than is possible with the use of the polysulfone alone. For example, in the next section of this chapter, blends of polysulfone with other high temperature polymers such as polyimides (PIs) and polybenzimidazole (PBI) will be discussed. Much of that effort is focused on the production of miscible blends that can be fabricated into both symmetric and asymmetric membranes. Later sections of this chapter will focus on the use of polysulfone in mixtures to modify other properties of polymers, particularly the fracture and impact behaviors.

6.3 Issues in blending polysulfone with other high temperature polymers

Following up on the comments from the last section, the first blends to be discussed will be blends of PBI with polysulfone. In an initial investigation of these blends, Chung *et al.*[1] determined that the two polymers do not form miscible mixtures. However, when fibers were spun from the blends, the tensile modulus and strength of the blend fibers were comparable to or better than that of PBI, depending on the processing conditions that were used.

It was later shown[2] that the introduction of functional groups, such as sulfonate groups, into the polysulfone polymer entity resulted in the formation of miscible blends with PBI. Further, the degree of sulfonation as well as the overall blend composition controls the observed miscibility behavior. Fourier Transform (FT-IR) Analysis showed that the observed miscibility is due to specific interactions between the N-H group on the PBI and the sulfonate groups on the polysulfone.

Similar observations have been made of blends of polyamide 11 (PA 11) and sulfonated polysulfone (SPSF) prepared by solution casting from dimethyl formamide (DMF).[3] In that work, differential scanning calorimetry (DSC) showed a melting point depression of the equilibrium melting point of the PA 11, indicative of an interaction between the two polymers. With lower degrees of sulfonation of the polysulfone, less interaction between the two polymers was observed. FT-IR and Raman spectroscopic techniques were used to confirm the nature of the specific interactions involved.

In EP 0778077A3, gas separation polymer membranes were prepared from mixtures of polysulfone, Udel P-1700, and an aromatic polyimide, Matrimid 5218.[4] The two polymers were proven to be completely miscible as confirmed by optical microscopy, glass transition temperature values and spectroscopy analysis of the prepared mixtures. This complete miscibility allowed for the preparation of both symmetric and asymmetric blend membranes in any proportion from 1 to 99 wt. % of polysulfone and polyimide. The blend membranes showed significant permeability improvements, compared to pure polyimides, with a minor change in their selectivity. Blend membranes were also more resistant to the plasticization phenomenon compared with pure polyimides. This work showed the use of polysulfone-polyimide blends for the preparation of gas separation membranes for application in the separation of industrial gases.

In another development that was focused on membranes, miscibility in blends of polysulfone with poly (1-vinyl pyrrolidone-co-styrene) (P (VP-S)) copolymers containing various amounts of 1-vinyl pyrrolidone (VP) was examined.[5] Copolymers that contained VP from 68 to 88 wt. % were used in that work and were found to be miscible with polysulfone. On the other hand, polysulfone blends with P(VP-S) copolymers containing 65 wt. % VP and those with the P (VP-S) copolymers containing 90 wt. % VP showed two T_g values, indicating that phase separation had occurred. In terms of membrane performance, the solute rejection examined with membranes fabricated from miscible blends was similar to that of a polysulfone membrane, while the solute rejection examined with the membranes fabricated from immiscible blends was lower than that of polysulfone membranes.

6.4 Physical properties of polysulfone blends

The previous section addressed miscibility issues encountered in polysulfone blends and how those relate to membrane performance. However, the functioning as a membrane is just one aspect of parts fabricated from polysulfone blends. This section of the chapter will deal with other properties, primarily mechanical properties.

Kohlman and Petrie[6] produced blends of 5%, 10% and 20% by weight of polysulfone in polycarbonate by melt blending. The resulting blends were injection molded into plaques. Mechanical analysis was performed on samples machined from the plaques and consisted of tensile, pendulum impact and ballistic impact testing. The average impact strength and the percentage of ductile failures decreased with increasing amounts of polysulfone. Tensile test results showed that a relationship exists between the percent composition and the yield strength for the molded blends, with the blends showing an improvement in tensile strength. The ballistic testing results showed that a linear relationship exists between the percent polysulfone in the blend and the critical velocity for complete penetration.

In another study,[7] Garcia *et al.* produced blends of polysulfone with a liquid crystal copolyester, Rodrun 5000, by processing methods of direct injection molding and extrusion followed by subsequent injection molding. The blends were immiscible and showed two pure polymer phases. A generally linear relationship of the Young's modulus of the directly molded blends was observed and was attributed to the counteracting effects of the large orientation of the molded bar skin and its low thickness. There is an improvement in the notched impact strength of the polysulfone with the addition of small amounts of the LCP that indicates a reduction in its notch sensitivity. The behavior of the tensile strength was close to linear with respect to the blend composition, except for a 20/80 blend which showed synergistic behavior. This behavior is reminiscent of what is sometimes observed in rubber-toughened blends.

Ramiro *et al.*[8] studied the mechanical properties of directly injection molded specimens based on blends of poly (etherimide) (PEI) and polysulfone. The molded samples were composed of an almost pure PEI phase and a pure polysulfone phase. The measured modulus of elasticity was near to an additive relationship of the two component polymer values. The impact strength was higher than that of the poly (etherimide). The tensile strength and elongation at break were almost additive, suggesting that there is good interfacial adhesion between the two phases. The reported behavior of the physical properties is again somewhat reminiscent of rubber-modified blends.

Polysulfone/PET blends were obtained by direct injection molding across the entire composition range.[9] A slight synergistic behavior was observed in the Young's modulus, mainly in the 90/10 blend. This is probably due to orientation effects. The measured ductility values were approximately linear with composition. This was generally the case with the polysulfone-rich blends and was attributed to some polysulfone being present in the PET-rich phase. The presence of some broken particles suggested some interfacial adhesion being present.

Summarizing the physical property results, then, on polysulfone blends, it seems that the strength and modulus values can be given by a linear relationship if the proper blend morphology is established. At the same time, it appears to be possible to observe synergistic behaviors in the impact properties for certain blend formulations. These results suggest that a toughening mechanism effect may be occurring in some of these polysulfone blends. That effect will be explored further in the next section of this chapter, when a discussion of polysulfone/thermoset mixtures is provided.

6.5 Polysulfone/thermoset mixtures

It is well recognized that phase separation is a necessary condition for improving the fracture toughness in thermoplastic-modified thermosetting matrices.[10,11] For a certain thermoplastic-modified thermosetting matrix, two primary factors control the final morphology and, hence, the ultimate properties of the mixtures: thermodynamics and the kinetics of phase separation during the curing process itself. Both these factors can be modified by changing the thermoplastic content in the mixture and/or varying the curing conditions.

Martinez *et al.*[12] performed such a study of polysulfone-modified diaminodiphenyl methane-cured diglycidyl ether of bisphenol-A epoxy mixtures. The immiscibility for the mixtures was proven for various polysulfone contents and as a function of the curing conditions used. It was found that the control of the general morphologies can be performed only by varying the processing temperature. For a particular thermoplastic amount, it was discovered that immiscibility lowers the rigidity and the strength but increases the fracture toughness of the mixtures. Those properties, including the fracture toughness, are also dependent upon the percentage of thermoplastic and the subsequent morphologies. The higher values of fracture toughness are achieved for a bicontinuous morphology.

Hwang *et al.*[13] have investigated mixtures of polysulfone with bisphenol A dicyanate, the structure of which is shown in Fig. 6.2. When blends with

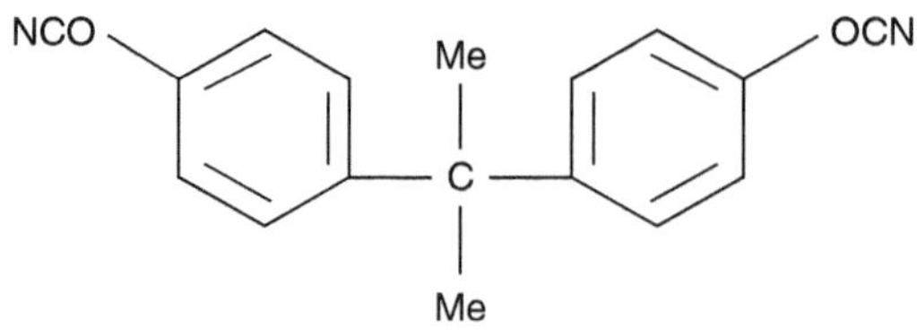

6.2 Chemical structure of bisphenol A dicyanate.

less than 10 wt. % of bisphenol A dicyanate were cured isothermally, they were phase separated due to a nucleation and growth mechanism. On the other hand, with more than 20 wt. % polysulfone, the blends were phase separated by spinodal decomposition. These different mechanisms for the phase separation are important because in thermoset/thermoplastic blending, the fracture toughness is determined by the morphology formed as a consequence of the phase separation.

Those preliminary studies were later extended to include the addition of organic montmorillonite nanoscale filler to the polysulfone/dicyanate mixtures.[14] The major reason for the introduction of the filler is the significant improvement in properties that can be obtained at low clay contents. The exfoliated configuration for the clay is of particular interest because it maximizes the interactions, making the entire surface of the clay layers available for interactions with the polymer matrix. This should lead to dramatic changes in mechanical and physical properties.

In the quoted study, it was found that the polysulfone is initially miscible with the cyanate prepolymer and phase separates in spherical microdomains during the course of the cyanate polymerization process. The flexural modulus was not significantly modified by the thermoplastic/organoclay addition. On the other hand, the fracture toughness was slightly improved with the addition of polysulfone and clay to the cyanate.

All of these studies highlight the importance of obtaining the correct morphology in the polysulfone/thermoset mixtures in order to be able to maximize the fracture toughness improvement observed. The mixtures often begin as miscible blends and, upon curing of the thermoset component, phase separation occurs. It is control of that phase separation process that ultimately leads to the attainment of a particular physical property profile. Variables that can affect the phase separation process include the overall mixture composition as well as the curing temperature of the thermoset component in the mixture. Among the blend composition variables, the molecular weight of the two components is very significant due to its effect on the overall mixture viscosity, an important factor in controlling the phase separation kinetics.

These conclusions imply that fundamental studies of the phase separation process in polysulfone/thermoset mixtures are needed to better understand the kinetics of the situation. General conclusions about the range of variables that allow for the development of a certain morphology compared to others are needed to be able to take full advantage of the fracture toughness enhancements that have been observed in certain polysulfone/thermoset mixtures. It is through such an understanding that the optimum polysulfone/thermoset mixture can be defined and developed.

Variables that should be examined in these type of studies are the molecular weights of the two component materials, both the thermoplastic and thermoset components. It should be determined if both of these affect the viscosity in similar ways, or if it is not the overall viscosity that affects the phase separation kinetics but how the overall mixture viscosity is achieved. Those studies should be done on a single system as a function of the curing temperature, and a complete understanding of the resulting morphologies should be obtained. Having obtained that basic and fundamental understanding on a single system, the goal would be to apply that knowledge to be able to design the required features of other systems. That way, systems with particular and specific property profiles can be designed and fabricated.

6.6 Conclusions

There are two general application areas for which polysulfone blends are currently being utilized. The first is the general development of high temperature separation membranes that use the membrane features of polysulfone in combination with the high temperature performance provided by the other polymer in the mixture. In general, miscibility of the two polymers is desired in this application, and much work has already been devoted to the development of miscible polymer mixtures. As with most polymer blends, one of the key unanswered questions on this topic is what are the key polymer features that lead to miscibility. Much effort has been provided to a trial-and-error approach in this regard to define miscible polymer pairs. It would be quite desirable to establish the general rules of miscibility and better define the relationship between miscible mixtures and membrane performance.

The other general application of polysulfone blends is in the area of impact-modified materials. The general approach adopted here has proven to be useful when the polysulfone is blended with both thermoplastic and thermoset matrix polymers. In this case, immiscibility of the two polymers is the desired phase structure, but control of the phase separation process is needed to be able to optimize the enhancement in fracture and impact properties of the matrix polymer. In other words, there is more to the observed effect than simply blending the two polymers together, but an understanding of the effect of the mixing parameters on the subsequent blend morphology needs to be obtained.

This leads to the one area in which additional research work in polysulfone blends is necessary. That area revolves around the general need to better understand the phase structure of polysulfone blends and how a particular phase structure can be obtained.

Polymer features that lead to miscibility with polysulfone should be further quantified to be able to optimize the membrane separation characteristics of polymer mixtures. On the other hand, in the case of immiscible polysulfone blends, it is desirable to better define the features of the blend components that lead to a particular morphology. Some of those features are perhaps going to be different in the case of thermoplastic and thermoset matrix materials, but viscosity is certainly going to be relevant in both cases. However, in order to best utilize the polysulfone blends that have been discussed in this chapter, more work is required to better comprehend their structure–property–processing relationships.

6.7 Sources of further information and advice

No books solely focused on polysulfone polymers or blends of polysulfones are available. A good general source of information on many of the general features of polysulfones is the Society of Plastics Engineers (SPE) encyclopedia.[15] There, information about many of the chemical and physical properties can be found.

6.8 References

1. T.S. Chung, M. Glick and E.J. Powers, *Polym. Eng. Sci.*, **33** (16) 1042 (1993).
2. V. Deimede, A. Voyiatzis, J.K. Kallitsis, L. Qingfeng and N.J. Bjerrum, *Macromolecules*, **33** (20) 7609 (2000).
3. V.A. Deimede, K.V. Frajou, J.K. Kallitsis and G.A. Voyiatzis, *Polymer*, **41** (26) 9095 (2000).
4. X.S. Dabou *et al.*, EP 0778077A3, Filed Dec. 5, 1996.
5. J.H. Kim, J.E. Yoo and C.K. Kim, *Macromol. Res.*, **10** (4) 209 (2002).
6. W.G. Kohlman and S.P. Petrie, *Adv. Polym, Tech.*, **14** (2) 111 (1995).
7. M. Garcia, J.I. Eguiazabal and J. Nazabal, *Polym. Intern.*, **53** (3) 272 (2004).
8. J. Ramiro, J.I. Eguiazabal and J. Nazabal, *Macromol Mater. Eng.*, **292** (10) 1059 (2007).
9. A. Retolaza, J.I. Eguiazabal and J. Nazabal, J. *Appl. Polym. Sci.*, **93** (5) 2193 (2004).
10. C.B. Bucknall and I. Partridge, *Polymer*, **24** 639 (1983).
11. J.N. Hay, B. Woodfine and M. Davies, *High Perform. Polym.*, **8** 35 (1996).
12. I. Martinez, M.D. Martin, A. Eceiza, P. Oyanguren and I. Mondragon, *Polymer*, **41** 1027 (2000).

13. J.W Hwang, S.D. Park, K. Cho, J.K. Kim, C.E. Park and T.S. Oh, *Polymer*, **38** (8) 1835 (1997).
14. I. Mondragon, L. Solar, A. Nohales, C.I. Vallo and C.M. Gomez, *Polymer*, **47** 3401 (2006).
15. *SPE Online Plastics Encyclopedia*, section on Polysulfones. Available at http://speplasticsindustryresource.com/

7

Polybenzimidazole (PBI) high temperature polymers and blends

B. G. DAWKINS, F. QIN, M. GRUENDER and G. S. COPELAND, PBI Performance Products, Inc., USA

DOI: 10.1533/9780857099013.174

Abstract: This chapter discusses polybenzimidazole (PBI), PBI blends, and some of their high temperature applications. The chapter first reviews the PBI polymer, including its history, structure, properties, synthesis, advantages and limitations. It then focuses on various PBI blends and their properties. The chapter also summarizes current commercial PBI and PBI blend products and provides examples of PBI and PBI blends in high temperature applications.

Key words: polybenzimidazole, high performance polymer blend, high temperature applications.

7.1 Introduction

Development work on aromatic polybenzimidazole (PBI) polymer was done in the early 1960s.[1,2] Later, NASA and the Air Force Materials Lab sponsored work on PBIs as a non-flammable and thermally stable textile fiber for aerospace and defense applications. In the 1970s NASA used PBI as part of the astronaut's clothing on Apollo and other space shuttle flights. In 1983 the Celanese Corporation commercialized PBI fibers, spun from solutions of the polymer, and subsequently started development of other PBI forms, including films, fibrids, papers, microporous resins, sizings, coatings, molding resins, as well as reinforced composites. PBI was introduced to the Fire Service as an outer shell protective fabric, typically 40% PBI/60% para-aramid. In the 1990s shortcut PBI fiber was introduced for use in automotive braking systems as a friction formulation material. Also, staple fiber was introduced into the aircraft market as a seat fire blocking layer material, and lightweight fabrics were developed for electric utility and petrochemical applications. During the early 2000s enhanced and next-generation fibers were continuously commercialized and introduced in firefighter turnout gear. In 2005, Celanese sold its PBI business to PBI Performance Products

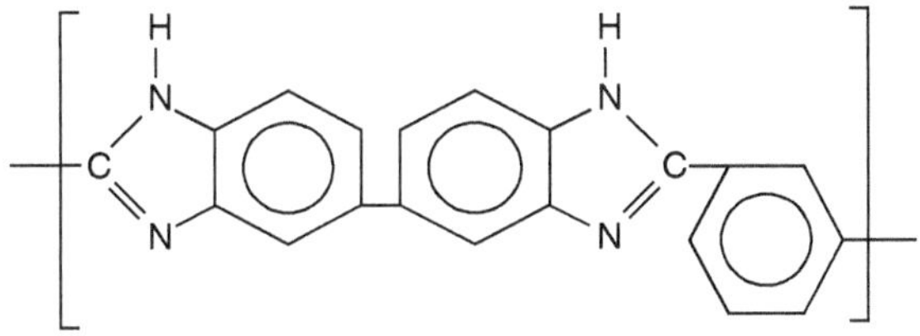

7.1 Structure of PBI polymer.

Inc, an affiliate of the InterTech Group of North Charleston, SC, and the world's only commercial producer of PBI polymer (Fig. 7.1) and fiber.

Thermal gravimetric analysis (TGA) of PBI polymer shows that decomposition in air occurs at around 580°C and, for short times at temperature, the polymer can withstand high temperatures without severe loss of mechanical properties. PBI does not burn in air, has high heat resistance (does not melt), possesses the highest compressive strength and mechanical property retention over 205°C of any unfilled resin, and is hydrolytically stable to high pressure steam or boiling water. In an inert atmosphere, such as nitrogen, the high temperature weight loss occurs at 600°C. For example, for a ~10-min exposure time, the suggested maximum temperature is ~650°C. Long-term temperature aging can lead to rapid loss of mechanical properties, due to oxidative degradation, and thus ~320°C is the maximum temperature capability for exposures exceeding 200 h. Under cryogenic conditions, it has been reported[3,4] that PBI maintains its properties in temperatures as low as minus 196°C. PBI is typically resistant to organic acids, chlorinated solvents, alcohols, and weak organic bases, but can be affected by polar aprotic solvents and strong acids and bases. PBI is also known to be hydroscopic with an equilibrium moisture content of ~15% by weight, the uptake primarily due to water forming hydrogen bonds with the hydrogen on the imidazole ring of the polymer chain. Thus, sample materials are generally dried at ~180°C for several hours in a vacuum oven prior to testing and usage. The polymer also absorbs acid (pKa ~ 5.5), a vital characteristic for fuel cell membranes and other proton-conducting applications. Figures 7.2 and 7.3 show the TGAs in nitrogen and air, respectively, at a rate of 20°C/min.

Due to the several attractive properties of PBI such as commercial availability, high glass transition temperature (T_g, 425–435°C), chemical resistance in hostile environments, and retention of good mechanical properties at both high and cryogenic temperatures, it has been and continues to be examined in high performance polymer blends.[5] Thus, blending with other high temperature-resistant polymers targets the high T_g provided by PBI and the improved processability afforded by the component polymers.

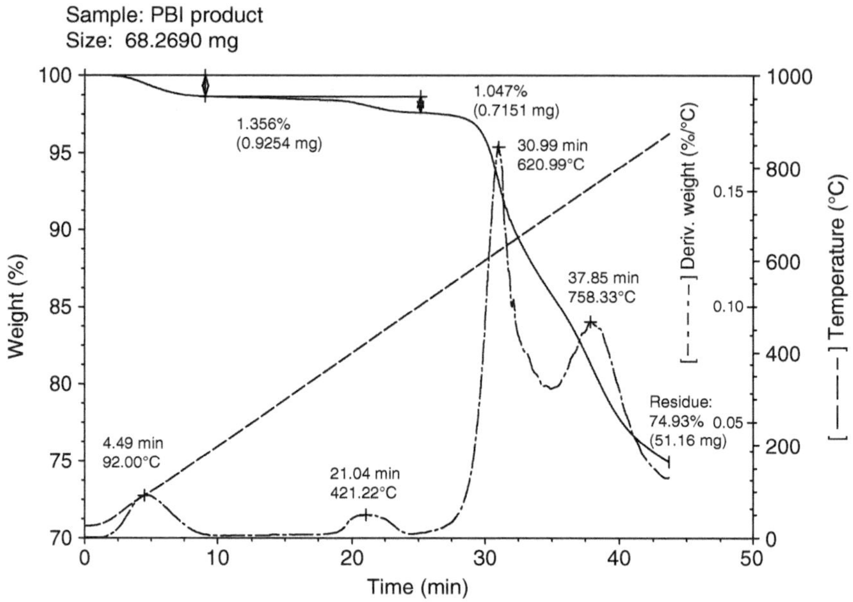

7.2 TGA curve of 100 mesh PBI (N_2).

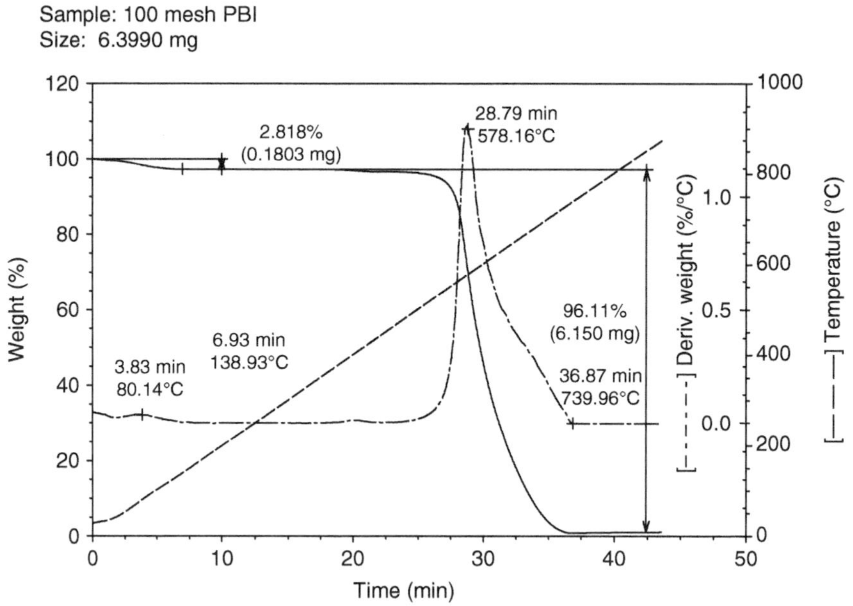

7.3 TGA curve of 100 mesh PBI (air).

High temperature-performance polymer needs often emerge where blend technology is required or expedites performance optimization of available materials. Ideally, the resulting blend is a miscible combination where phase separation does not occur during thermoplastic processing.

This chapter covers PBI-based polymer blends, their preparation, their unique properties relative to the companion polymer, and their applications or potential applications. Typically, these blends would consist of more than one polymer, specifically having high performance components or resulting in high performance materials, as opposed to new chemistry to produce new polymer materials. Blending allows for new and unique property tailoring and with almost infinite possibilities, and within targeted performance and price specifications. Such blends take advantage of the high T_g of PBI (425–435°C) relative to the component polymers, the blends being melt processable at high temperatures, miscible or partially miscible, compatible, and applicable to high temperature end uses. With a resulting single T_g polymer blend, specific modifications of processing temperatures may be made, and with adherence to 'Rule of Mixtures' type predictable behavior and property ranges.

The chapter is intended to be a general overview, not totally comprehensive, but with coverage of specific and significant polymer blends systems that address specific deficiencies and needs based on current applications and trends in the field of high temperature polymer blends, particularly with PBI. PBI blends, synthesis, structures, modified structures properties, development activities, and applications were reviewed and covered previously.[6–11] Much of the previous work[7] emphasized the use of polymer materials in application areas traditionally dominated by metals and alloys, such developments including new polymers, control of morphology, reinforcements, and blending, and with blending receiving less attention for applications in the 300–400°C range. Subsequently, much effort has been focused on the preparation and characterization of high performance polymer blends toward developing a better fundamental understanding of the phase behavior of such blends, and end-use potential.

7.2 Processing of polybenzimidazole (PBI)

As a class of polymers, polybenzimidazoles exhibit high thermal and chemical stability. PBI is unique as a commercially available over 400°C T_g resin, but lacks the thermoplastic processability and has relatively poor long-term thermo-oxidative stability, as well as relatively high moisture regain (at 10–15%) vs. polyimides, polycarbonates, and polyarylketones (up to 2%); the regain is apparently due to intermolecular hydrogen bonding between water and N-H groups in the PBI structure. PBIs can be further processed into various product and end-use forms such as fiber, solutions, coatings, molded shapes, film, membranes, foams, and composites. Additionally, blends with other polymer resins and materials can be processed to take advantage of the unique high performance properties, modified for desired thermal, chemical, processability, and mechanical targets, and end-use applications. The thermal stability of polybenzimidazoles is inherent in the wholly aromatic, ladder-like structure (Fig. 7.1), with the presence of three benzene rings in the repeating unit contributing to the polymer's superior thermal stability properties, stiffness, and toughness. Blending helps to improve processability since the polymer typically does not melt, and variant polymer properties can be achieved, as opposed to varying/tuning the polymer structure and subsequent properties.

PBI polymer is manufactured exclusively in commercial quantities by PBI Performance Products, Inc. The fiber processed from PBI polymer solutions is recognized as high performance for its exceptional thermal stability and chemical resistance. The commercial process utilizes two commercially available monomers, tetraaminobiphenyl (TAB) and diphenylisophthalate (DPIP), in a two-stage melt condensation and solid-state polymerization (SSP) process. The monomers are heated in an inert atmosphere. As the reaction proceeds in the first stage, PBI prepolymer is formed, and water and phenol foaming by-products evolve. The reactor is cooled and the prepolymer crushed. In the second stage reaction, the prepolymer is heated additionally at higher temperature to complete the polymerization and advance the molecular weight to the target value for further processing.

The reactor is cooled and poly(2,2′-(m-phenylene)-5,5′-bibenzimidazole), the only commercially available PBI polymer (Fig. 7.1), is discharged. The polymer is tan/golden brown in color. This particular commercial form of PBI is available in polymer resin form, under the trade name Celazole®, and is also available in solution form (~26%, 10% and other levels of polymer in N,N′-dimethyl acetamide (DMAc), or other solvents). PBI polymer of 100-mesh particle size is typically used in the manufacture of dry blends for extruded and injection molded stock shapes. Polymerization can be carried out either in solution or under molten states with various monomers, i.e., an aromatic tetraamine and a diphenyl ester or an anhydride of an aromatic or heterocyclic dicarboxylic acid in a one- or two-stage process. The commercial PBI product is meta-PBI where the phenylene ring is meta-coordinated. The para-PBI polymer can provide superior tensile strength and stiffness, but is not as soluble in useful solvents for downstream processing. Figure 7.4 shows a typical PBI polymerization process reaction.

A great deal of literature on PBI has been developed since the first reported synthesis.[7,12–14] Many aromatic PBI variants have been synthesized and studied, but the commercial PBI product is the one cited most often and receiving most attention in developmental applications; it is particularly

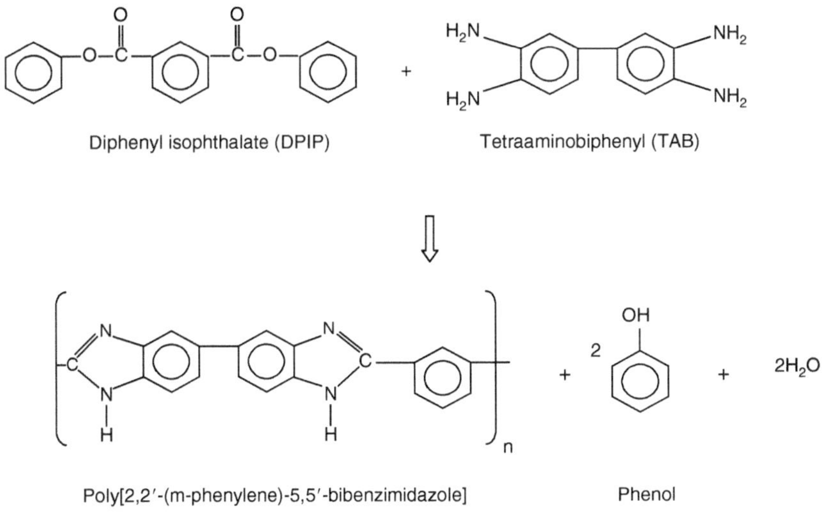

7.4 PBI polymerization.

used as a polymeric matrix material capable of maintaining load bearing properties for short periods of time at temperatures up to ~650°C, and with good mechanical properties at elevated temperatures, good chemical resistance, and potentially useful in a variety of aerospace applications.[15] Aromatic PBIs are characterized by a high degree of thermal stability. They may be shaped to form fibers, films, and other articles of wide utility which show resistance to degradation by heat, hydrolytic media, and oxidizing media. However, these PBIs are not readily melt processable at desirable low temperatures and pressures, and so shaped articles are not readily fabricated from them. Molded parts, for example, require a two-stage process involving cold compaction for the PBI powders, followed by heating at pressures of 1–3 Kpsi at 425–500°C for a few hours.[16]

PBI polymer solutions are prepared by dissolving the polymer resin in dimethyl acetamide (DMAc)/LiCl (lithium chloride) under pressure and temperature in an inert atmosphere. The LiCl is added to stabilize the solution, increasing the shelf-life from days to several months. Prior to further processing, the PBI dope solution is filtered to remove insolubles that could interfere with down-stream processing, such as spinning into fiber. Solutions may also be prepared without the added stabilizer, particularly at lower solids levels and for applications requiring the absence of metal or interfering additives.

7.3 PBI blends

Polymer blending allows the opportunity to lower costs without sacrificing the desired properties, the ability to tailor properties without creating completely new polymers, and presents options for high performance blends from synergistically interacting polymer systems. Miscible polymer blends and co-polymerization can offer complementary routes to polymer systems of tailored properties, although co-polymerization can often heighten miscibility and lead to better routes to control cost-performance. The optimum blend system is often derived after evaluation of process ease and performance profiles available, which could involve blending, co-polymerization, or a combination.[6] Compared to polymerizations or co-polymerizations to form new polymers or variants, polymer resin blending is often the preferred route[17] due to simplicity, reproducibility, processability, and reduced development cost. Such blending work by Chung and others showed that incorporation of small amounts of PBI (3–17%) into Matrimid ® 5218 polyimide increased the thermal stability and selectivity of the blend film for pervaporation dehydration of t-butanol. Chung and Xu[18] described solution spinning of hollow fibers based on a miscible blend of PBI and Ultem ® 1010 polyetherimide (PEI). Furthermore, blending can synergistically combine the advantages of different material properties and overcome the deficiencies of the individual components, and miscible blends are desirable for producing final products with uniform performance and stable thermal and mechanical properties. Property targets can thus be tuned to deliver end-use attributes associated with T_g, melt processability, properties tailored to the required application as opposed to over-performance, solubility controls and with alternate solvents, cost controls, i.e., for a confluence of properties and conditions to enter into a variety of different applications. Sudhir and Kratzer[19] also have reported fluid separation membranes made from blends of polyimides and polyimidazoles; such membranes are reported to exhibit resistance to interaction with the fluid stream material which can result in plasticizing of the membrane, have superior mechanical properties for performance in high membrane differential pressure and high process temperature environments, can be configured in hollow fiber membranes, and have good performance in long-term contact with aggressive process composition, pressure, and temperature conditions.

Polymer blends may be miscible, immiscible or partial (a combination of the two).[16] Miscible blends offer additional desirable and, often needed, end-use properties intermediate to the individual components across the miscible composition range. Typically, these have uniform physical properties. As stated by Flory:[20] 'It is well known that, regarding the mixing of thermoplastic polymers, incompatibility is the rule and miscibility or even partial miscibility is the exception. Since most thermoplastic polymers are immiscible with other thermoplastic polymers, the discovery of a homogeneous mixture or partially miscible mixture of two or more thermoplastic polymers is, indeed, usually unpredictable with any degree of certainty'. And, it is generally known or deduced that most pairs of polymers tend to be totally immiscible, even some of the same type.[21] Thus, miscible polymer blends are not common, but can be induced with temperature and relative composition considerations.

A miscible blend of two polymers will generally have properties somewhere between those of the two polymer components. The component with the higher property value, for example T_g, will increase that value of the blend as the relative amount of that component is increased, generally linearly, but deviations may occur if the two polymer components bind more strongly to each other than to themselves, if intermolecular bonding is involved, etc., since stronger binding leads to decreased mobility. However, in most cases, two polymers would be expected to bind less strongly with each other than with themselves. Other properties, such as mechanical properties, resistance to chemicals/radiation/heat, etc., would generally trend the same as T_g with respect to relative amounts of each polymer in the blend. Immiscible mixture or phase-separated blends will typically display a T_g corresponding to each phase, or a T_g for each component if totally incompatible.

There are several technical and manufacturing reasons to blend polymers in various forms, including to facilitate processing, to improve or modify properties (mechanical, physical, chemical, absorbency, thermal and thermal oxidative stability, hydrolytic stability, chemical resistance), to reduce/control costs, etc. The resulting blends and targeted properties are typically performed in concert with product applications or new applications identification (Celanese internal communications). The latter is usually more long-term and encompasses more advanced planning, recognition of market needs, more testing, and eventually insight into specific performance needs at some reasonable price. Applications and end-use testing are usually already available or adaptable, often in the form of standard protocols, whereas this may or may not be the case for new applications or new specialty forms of existing base polymers and materials to meet a needed/required product-market situation. Besides blending, other options to tune polymer properties include sieving to isolate specific particle sizes or molecular weights, cross-linking polymer structures, cross-linking blends, modifying the polymer chain, as well as co-polymerization and post-polymerization treatments. PBI is potentially a good candidate for nucleophilic substitution reactions due to the N-H bonds in the structure.

Typical blend polymers for PBI include poly (aryl ether ketone) (PAEK), polyetherimide (PEI), poly (aryl ether sulfone), poly (phenylene sulfide), and mixtures thereof. Different types of PAEK resins (Fig. 7.5) include PEK, PEEK, PEKEKK, PEKK, PEEKK, etc., several of which have melting points over 300°C. PAEK resins with relatively low T_g values can be further improved by the addition of reinforcement fillers, such as fiberglass, carbon fibers, and ceramic or mineral fillers. Such additions, however, can

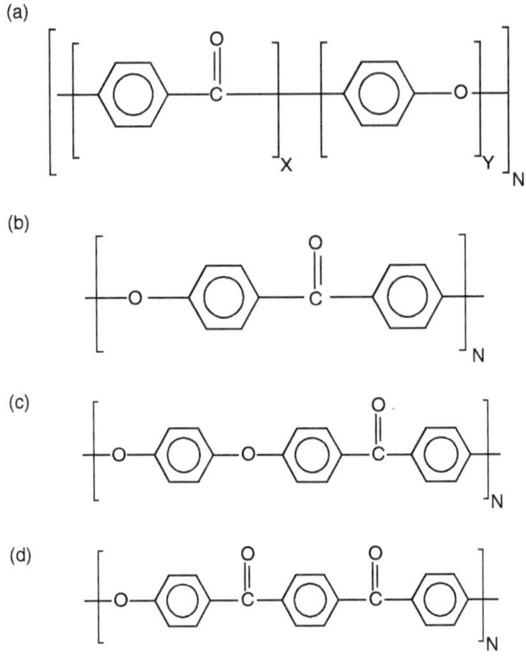

7.5 Polylaryletherketone structures. In panel (a), X, Y and N are positive integers. There are several variants of PAEK, for example: (b) polyetherketone (PEK); (c) polyetheretherketone (PEEK); (d) polyetherketoneketone (PEKK).

compromise or be neutral in terms of material weight, creep resistance and relaxation under stress. The predominant factor for high performance applications is T_g, and the use of blending to increase or synergize. PBI resin can be utilized to increase the T_g of the blend, or the other blend component used to provide a lower or intermediate T_g for intermediate applications. PBI is not expected to be detrimental to the other blend component as the other polymer usually becomes the limiting factor, although the other polymer can enhance melt processability of the blend.

Due to its very unique and high thermal stability characteristics, PBI blends were studied and developed early on for aerospace and composite materials.[22] The discovery of new polymeric alloys by blending commercially available polymers was the trend in materials development. A key reason for growth in this area was the ability to tailor existing materials to a new/unique set of property-performance-price specifications through combinations of materials. For example, a miscible blend based on PBI and polyimide (PI) was discovered and, after thorough investigations, PBI was found not only to have superior thermal/chemical properties, but to also exhibit unique reactivity to enable it to interact with other functionalized polar molecules to form compatible binary systems.[23] PBIs reactivity also led to the discovery of interactions between PBI and polyarylate (PA) polymers at Hoechst Celanese, which resulted in the preparation and characterization of PBI/PA film and fiber blends. At a fixed concentration (e.g. 0.5%), it was observed that the inherent viscosity (IV) of the solution blends in N-methyl-2-pyrrolidone (NMP) exceeded the Rule of Mixtures, suggesting PBI and PA exhibited interaction in a dilute solution such that the resulting hydrodynamic sizes of the blends was greater than the calculated averages based on each component. This interaction was further demonstrated by FTIR; the carbonyl stretching shifted downfield, indicating the existence of intermolecular H-bonding between PBI and PA in a film blend. Thus, a new family of high performance, processable, and compatible blends based on PBI and PA were discovered, which exhibited synergistic effects on properties and offered new opportunities for polymer alloys for engineering and aerospace applications.

High temperature polymer blends are typically prepared to take advantage of the high T_g, such as PBI with polyaryletherketones (PAEKs), typically polyetheretherketones (PEEKs) and polyetherketoneketones (PEKKs). PAEKs render PBI more melt processable and provide synergism for varying the end-use thermal and mechanical properties. The miscibility of such blends may be confirmed by the presence of a single glass transition temperature lying between those of the blend components.

7.4 PBI–polyetherketoneketone (PEKK) blends

Celazole® PBI is the low to medium molecular weight form of PBI molding resin, 23 000–37 000 grams per mole, inherent viscosity of 0.55–0.8 dL/g.[5] PBI and polyetherketoneketone (PEKK) can be solution blended with blend miscibility over a range of concentration ratios. Targeted blends may lower the moisture absorption vs. PBI, improve tractability vs. PBI, improve strength vs. PEKK, and with thermal and mechanical properties exceeding those predicted by the Rule of Mixtures. Polyarylketones (PAK) have good chemical resistance and moderate compressive strength, but exhibit poorer mechanical properties at elevated temperatures as compared to PBI. There are processing limitations to molding PBI articles, whereas PAK molded articles are limited in thermal and pressure resistance.

Blend patent work[24] describes a process for producing a solution blend of PBI and an amorphous PEKK. The objective was to prepare solution-derived blends of 100-mesh PBI polymer with PEKK, and to evaluate the thermal properties (Figs 7.6 and 7.7 show the TGA curves for PBI and PEKK, respectively). The PBI is mixed with sulfuric acid at a temperature between 40°C and 80°C for 30 min to 2 h to produce a PBI solution, then cooled to

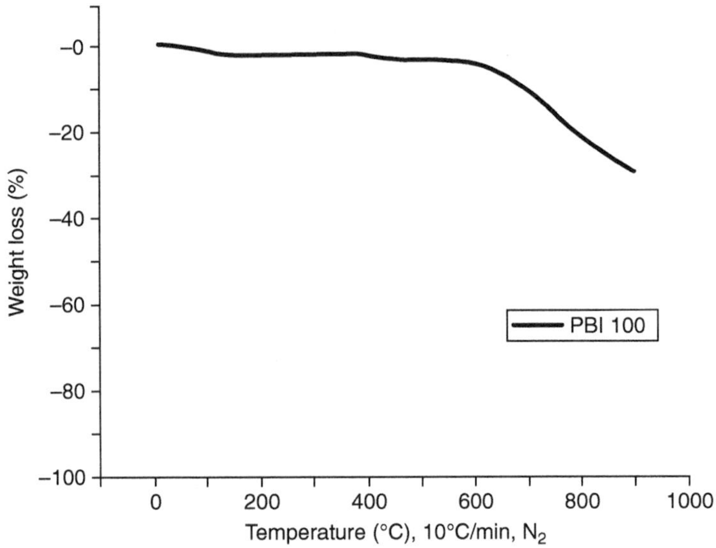

7.6 TGA curve of PBI 100.

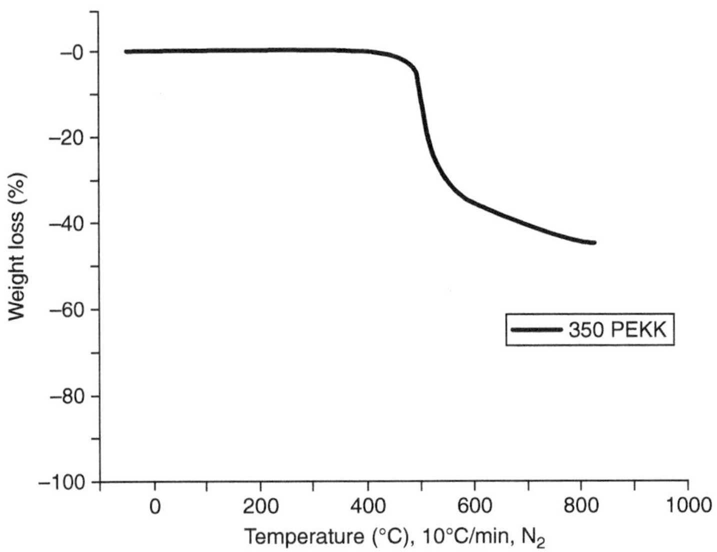

7.7 TGA curve of PEKK.

room temperature. PEKK is then added to the PBI solution to form a mixture, the mixture is stirred to homogenize, and subsequently poured into an excess of water, methanol, or water/methanol, being stirred swiftly. The aqueous mixture is filtered to produce a blend, and the polymer blend is washed with water and dried. Thus, the common sulfuric acid solvent can be used to form blends in varying ratios, then added to a non-solvent to isolate and recover the macroscopically homogeneous mixture.

Miscible blends of PBI and PEKK having a single or predominant T_g were found to be feasible from 50/50 to 10/90 (PBI/PEKK) with intimate and uniform mixing. While blends 40/60 to 90/10 may also be miscible, they tend to behave thermally more like PEKK (T_g ~ 150°C, amorphous phase; T_m ~ 335°C, melting point of the crystalline phase) vs. PBI (T_g ~ 430°C). There appeared to be a tendency to lose T_g, and thus potentially heat deflection temperature (HDT), as the PEKK ratio exceeded 40% (Fig. 7.8). The most compatible blends appeared to be 90/10 and 80/20. The amorphous PEKK with greater than 50% PEKK appeared to degrade as the pure PEKK component, whereas the blend with less than 50% PEKK appeared to degrade similarly to pure Celazole® PBI. As expected, the degradation temperature increased toward that of PBI as its content was increased (Fig. 7.9). The semi-crystalline blend containing greater than 30% PEKK appeared to degrade as the homopolymer PEKK, presumably due to the phase-separated domains of the crystalline segments.

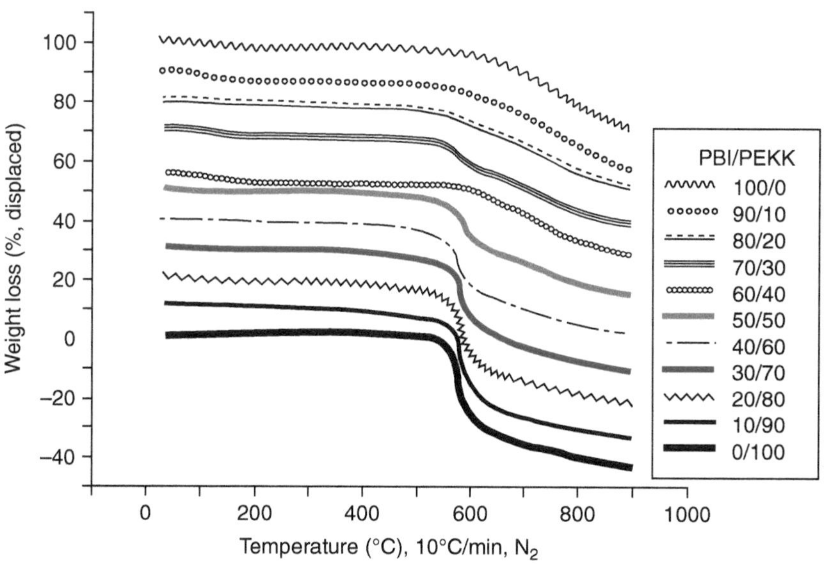

7.8 TGA curve of PBI/PEKK blends. Curves are displaced by 10 unit increment from bottom up.

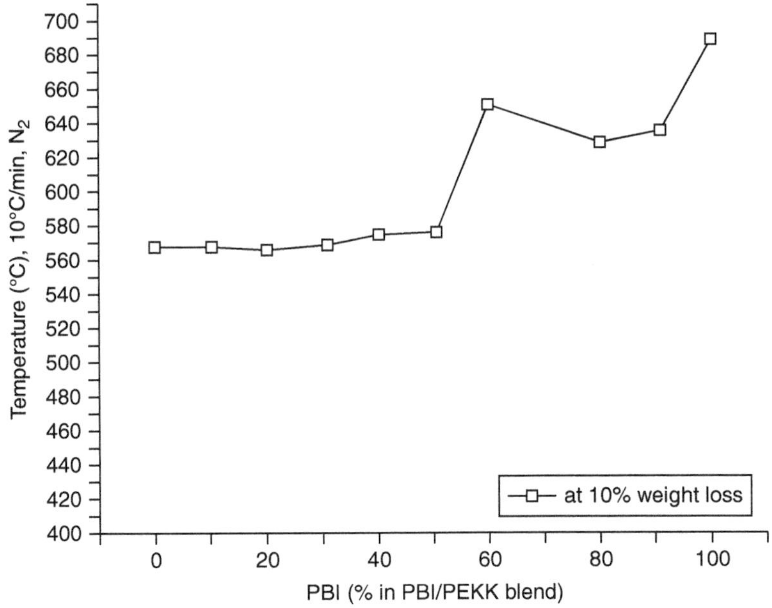

7.9 Temperature at 10% weight loss as a function of PBI content.

This referenced patent also includes a process for producing a melt blend of PBI and PEKK. In this alternative process, the PBI and PEKK are pre-mixed dry to obtain a dry mix. Then, in an extruder with a plurality of heating zones (240–410°C), the dry mix is fed and melted as it passes through the extruder, either single- or twin-screw, and is subjected to the mixing and the temperature generated; the resulting product is a melt blend of PBI–PEKK feasible in varying proportions from 1/99 to 80/20 PBI–PEKK.

PBI powder resin (Celazole® 100-mesh polymer) is typically chard-like in shape. When melt blended, SEM indicates that particle morphology is retained after compounding. Thus, PBI is not dissolving in the PAEK; it is apparently behaving as an encapsulating filler. It appears that the PBI polymer resin immobilizes the lower-melting-point amorphous regions of the PAEK matrix resin, thus imparting enhanced thermo-mechanical properties.

Unfilled grades of partially crystalline polymers such as PEEK are not very useful in molded part forms at temperatures much above their T_g since the softening that occurs leads to a sharp loss in polymer mechanical properties, especially stiffness/modulus. This lowered modulus is reflected in reduced dimensional stability under stress at temperatures above T_g.

Polymer mixtures with PBI benefit from PBI's ability to immobilize the amorphous regions of the PAEK matrix resin, thereby imparting enhanced thermo-mechanical properties not obtainable with other fillers of PAEK, but which fall short of that obtained in a miscible blend of polymers.

The Fox equation relationship (Fig. 7.10) defines or describes the composition dependence of the T_g of ideal co-polymer or ideal polymer blends (or miscible blends of two polymers) in which no strong interactions are involved. The Fox equation typically is expected to lead to a lower value of T_g than would be given by a simple linear Rule of Mixtures (Blend Inherent Viscosity vs. % Component in the Solution Blends) and reflects the effective higher free volume or randomness due to the presence of two components in the mixture:

$$\frac{1}{T_g} = \frac{M_1}{T_g^1} + \frac{M_2}{T_g^2}$$

where M_1 and M_2 are the mass fractions, T_g^1 and T_g^2 are the glass transition temperatures (in Kelvin) of polymers 1 and 2.

One would predict that a blend of PBI resin with PEKK would exhibit thermal properties roughly following the Fox relationship, but

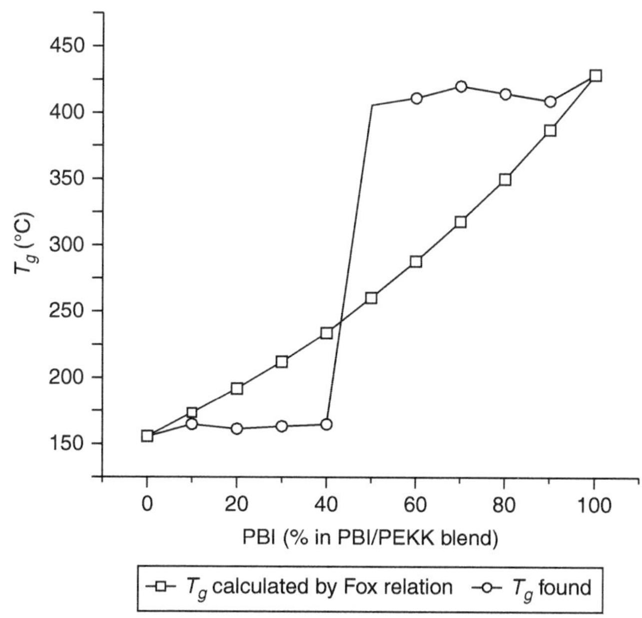

7.10 T_g values of PBI/PEKK blends as a function of PBI content.

unexpectedly it was determined that a PBI–PEKK blend in all proportions from 50/50 to 90/10 PBI–PEKK made by a solution process has a T_g of greater than 400°C, which is greater than would be predicted by the Fox relationship. Note also the TGA curves and thus thermal properties of the PBI and PEKK individual components, as well as for the various PBI-PEKK blends (Fig. 7.8).

The dried blend resin can either be injection or compression molded into articles with specifically tailored thermal and mechanical properties that lie between the respective properties of the pure PBI and PEKK from which the blend was produced. These blends can be used to enhance films, composites, or alloys. Further, these blends can be reinforced or filled prior to molding to produce unique blended products such as T-series products marketed by PBI Performance Products Inc.

7.5 PBI–polyetherimide (PEI) blends

PEIs (polyetherimides, Fig. 7.11) are basically combination polymers, consisting of both polyimide and polyether units in the backbone. Ultem PEIs, with T_g values in the 200–280°C range, are made on a commercial scale as engineering thermoplastics. Compared to the stiffer polyimides, PEIs can typically be melted and molded, presumably due to the ether linkages, but maintain good thermal stability, high strength, and high modulus.

PBI–PEI blends can be useful in severe environmental conditions and have good impact resistance, solvent resistance, and price/performance characteristics. PBIs are generally more expensive, have high moisture regain, but high T_g values and good solvent resistance. PEIs generally have lower T_g values, lack resistance to certain solvents, but are tractable and relatively less expensive.

Musto and others[25] reported FTIR investigation of the blend system consisting of PBI and poly (etherimide) (PEI, T_g ~ 220°C), poly [2,2′-bis[4-(3,4-dicarboxyphenoxy) phenyl] propane-m-benzenediamine], or commercially available Ultem® 1000, showing evidence of hydrogen bonding between

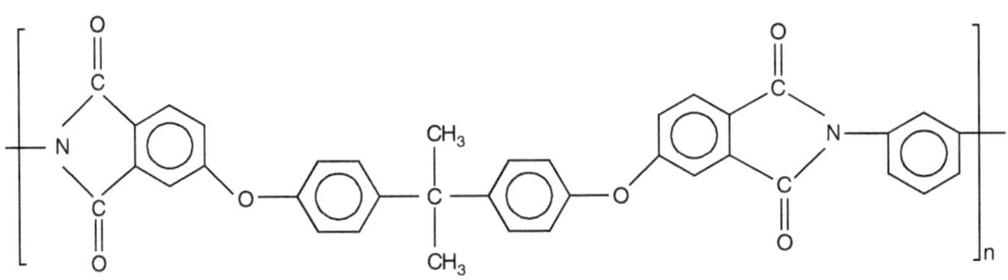

7.11 Polyetherimide structure.

the N-H groups of PBI and the imide carbonyl of Ultem® 1000. Such blends, solution cast from a common solvent, such as DMAc, DMF (N,N′-dimethylformamide), and NMP, were found to have a one composition dependent T_g. As-cast blends formed one-phase structures over the entire composition range;[26] however, optical microscopy and DSC (differential scanning calorimetry) revealed that the blends phase-separated upon heating above their T_g and did not remix upon cooling. The blends formed upon room temperature casting were non-equilibrium and remained one-phase because they were below T_g. The phase separation of such 50/50 blends was characterized by small-angle X-ray scattering and solid-state nuclear magnetic resonance,[27] showing separation upon annealing at temperatures of 310°C and above, although the DSC-determined T_g was reported at 344°C. Upon heating the blend above the blend T_g, the once-cooled blend was found to exhibit two T_g values, indicating phase separation at elevated temperatures. Phase separation in the blend is expected to be influenced by the vitrification of the PBI-rich phase owing to the much higher T_g of PBI (425–435°C) relative to that of PEI (~220°C). Blends of these materials are of interest because of the potential of obtaining materials with unique properties not available with either material alone. Such blends exhibit good resistance to chlorinated solvents,[7] lowered moisture uptake vs. PBI alone, and an intermediate T_g. Films of PBI, PEI, and PBI–PEI blends of all stoichiometries are soluble in DMAc.

Various compositions of blended aromatic PBIs and aromatic PEIs can be expected to have approximate linear T_g per weight % of PBI (Fig. 7.12) according to their phase diagrams based on single T_g values from single DSC scans on PBI–PEI blend compositions ranging from ~5% to 95% Ultem® 1000.[28]

The PBI amount in such blends should be at a level to render the composition less sensitive to solvents and to increase the T_g of the PEI. Blend compositions 1 to 99 parts by weight of PEI are feasible, but the 65–95% by weight of PBI blends appear to be the preferred ones. PEIs generally have good resistance to chemicals, with the exception being chlorinated solvents, ethylene glycol antifreeze, brake fluid and DMF, among others. Such blend compositions may be useful for molding, whereas PBI is not as useful.[28] By varying the nature of the diamine used to synthesize and the position of substitution in the bis-imides, and using different bis-phenols, a large number of aromatic PEIs with a wide range of physical properties can be prepared,[29] but blending with PBI offers non-synthesis routes to obtaining similar or improved properties. Cross-linking and addition of fillers and reinforcements may also be incorporated into such blends to improve and target specific property attributes.

Treatment of the PBI component, via acid protonation of the imidazole ring, should improve the thermo-oxidative stability of the PBI, as well as the

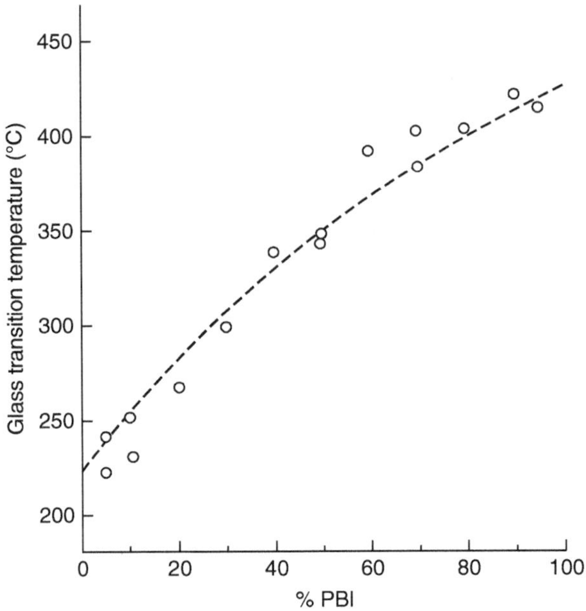

7.12 PBI/PEI blends, T_g vs. % PBI.

blend. Such blends are expected to be useful for matrix use up to ~260°C, and with good tensile and mechanical properties. Such blends can be useful as films, coating, or for molding. If the blend is mostly PBI, the PEI renders it more tractable and thermally processable, and less susceptible to moisture. If the blend is mostly PEI, the PBI increases the T_g and renders it more resistant to solvents. The composition is impact resistant and withstands severe environments. This same assessment would apply to PBI–PI blends.

7.6 PBI–polyaryletherketone (PAEK)–PEI blends

The targets for tri-blends consisting of PBI, polyaryletherketone (PAEK), and polyetherimide (PEI) include thermal and mechanical properties exceeding those of PAEK and PEI, lower moisture absorption vs. PBI, improved tractability vs. PBI, and improved strength vs. PAEK or PEI. The PAEK options include PEK, PEEK, PEEKK, PEKEKK, PEKK, etc. Some of the earlier work was performed toward reducing cost, crystallinity, and mold shrinkage, lowering warpage, enhancing PBI–PEEK interfacial interaction, modified blend rheology and improved matrix resin T_g, and a concern related to the thermal oxidative stability of PEI relative to improving that of PBI and the overall matrix.

Blend patent work[30] demonstrated tri-blends of PBI (T_g, 425–435°C), PAEK (T_g ~ 145°C), and PEI (T_g ~ 218°C), formed by intimately blending by dissolving in a strong acid or by melt blending. Various tri-blends were prepared, including 45/45/10, 37.5/37.5/25, 25/25/50, and 12/12.5/75 PBI/PEEK/PEI. It appeared that incorporation of 10 to 75 parts of Ultem 1000 PEI into a base mix of 50/50 PBI/PEEK resulted in a decrease in the degradation temperature. The incorporation of 10 to 25 parts PEEK into a base mix of PBI/PEI (50/50) resulted in a considerable decrease in degradation. The PBI/PEI blend prepared in sulfuric acid exhibited a higher thermal stability than that prepared in DMAc. The PBI/PEKK di-blend showed evidence of improved properties compared to the tri-blend.

DSC analyses indicated that the T_g of PEEK in the blend increased due to the formation of a miscible blend with the PEI. The crystallinity of PEEK was also found to decrease after mixing with the amorphous PEI. Thermal degradation of the PEI was found to be insignificant for the PEI-filled PBI–PEEK at the 10–35% level. Mechanical properties of PBI–PEEK (50/50) were found to be minimally affected with the addition of up to about 10% PEI. PBI and PEI have been demonstrated to be miscible and to exhibit strong molecular interactions.

PAEKs have good chemical resistance and moderate compressive strength, but poorer mechanical properties at elevated temperatures when compared to PBI. Blends of PBI and PAEK[31] and of PBI and PEI[28] have been reported, the former using a dry blend with a high speed mixer, and the latter using a mutual polar solvent such as DMAc, NMP, or DMF. These have good attributes, but further improvements are still desirable for some applications requiring enhanced properties.

PEIs typically have high heat resistance, high strength-to-weight ratio, high modulus, excellent non-flammability characteristics, processability on conventional molding equipment, low smoke evolution, high dielectric strength, a stable dielectric constant and dissipation factor over a wide range of temperatures and frequencies, and good chemical resistance, except to, for example, chlorinated solvents, ethylene glycol, and DMF.

The reported tri-blend composition contains a minimum of 10% by weight of each component and a maximum of 75% of any one component. The components are intimately blended and harvested. Blend solutions can be prepared by the use of a common solvent for each of the three components, such as sulfuric acid, methane sulfonic acid, trifluoromethane sulfonic acid, trifluoroacetic acid, dichloroacetic acid, and combinations thereof. Sulfuric acid was found to be the preferred solvent. With melt blending, the components can be intimately blended in an extruder. Prior to blending in the extruder, the components may be dry blended in, for example, a tumbler or high shear mixer. Harvesting, or blend recovery, can be performed by precipitating out the blend in a non-solvent such as water, methanol, or combinations thereof, and followed by filtering, washing, neutralizing any residual solvent, washing the neutralized tri-blend, and drying. When the blend is made by a melt blend process, the tri-blend may be ground to the appropriate particle size needed for further processing.

Applications of such tri-blends include semiconductor devices, including microchips, flat panel displays, and the like. In the manufacture of such apparatus, such blends can be used as components of the vapor deposition chamber, process housing components, and materials holding fixtures or molds, in order to be inert to the vapor deposition environment and to prevent off-gas contamination. These blends may also be used in components of semiconductor etching, ashing, wafer transportation, and as hard-disc media cassettes where support components may function as a seal, insulator, holding or transportation device which must endure exposure to the harsh processing environments. Further applications for such blends can be found in various industrial, chemical, and petrochemical processes where the temperature resistance in chemical environments extends the life cycle of parts used in such fabrication processes.

7.7 PBI–polyarylate (PA) blends

Polyarylates (PA) have the general structure $[\text{-OAOOCA'CO-}]_n$ where A is a phenyl radical and A′ is a carboxylic acid radical, wholly aromatic polyesters based on diphenols and aromatic dicarboxylic acids, or their derivatives. They have good processability in terms of injection moldability, high softening temperatures, and good thermal and mechanical properties for a variety of industrial applications, particularly those requiring high T_g values and good HDT. PAs are reasonably soluble in methylene chloride, chloroform, NMP, DMF, DMAc, DMSO (dimethylsulfoxide), and interact with and/or plasticize PBI for improved price/performance characteristics.

Several investigations have focused on a family of high performance blends of PBI and PA, as well as high modulus aramides (HMA).[22,32–34] IR spectra, transmission electron spectroscopy (TEM), and thermal mechanical analysis (TMA) showed that the degree of interaction depended on process conditions and LiCl stabilizer addition. Viscometry and FTIR measurements confirmed the existence of intermolecular hydrogen bonding between the two polymers in solution and in films cast from solutions. Solutions of PBI and a PA in 1-methyl-2-pyrrolidinone were found to be homogeneous and increased in stability with increasing PA concentration. The blend films were found to have significantly better solvent resistance than PA films alone. Phase separation can occur with removal of the solvent. The blend properties of PBI–PA and PBI–HMA fibers were found to vary with fiber aging, whereas 80/20 PBI–PS blend fibers showed no aging effects, i.e., no deterioration of mechanical properties.

Polyarylates are high temperature, high performance, thermoplastic polymers with good combinations of thermal and mechanical properties.

They also have good processability, which allows them to be molded into a variety of useful articles. They are, however, limited in the aerospace industry, where the chemical resistance is limited in chlorinated solvents and inorganic acids. Blends have been reported with polyesters,[35] as well as PEIs[36] and polycarbonates[37] and others. The property shortcomings of polyarylates can be substantially improved by incorporating PBI as a miscible blend.[38] Beneficial properties are reported to be maintained and solvent, resistance, thermal, and physical properties can be improved at a moderate additional cost.

Chung and others reported[38] on novel membranes comprised of a blend of polyarylate and PBI polymers, the blend membranes exhibiting enhanced properties over membranes prepared solely from either polymer alone. The addition of polyarylate allows the composition to be more thermally processable and less susceptible to moisture. The PBI renders the polyarylate less reactive to solvents and increases its thermal stability. Additionally, such blend membranes were found to show high regeneration capacity while retaining good flux ranges. Such membranes demonstrated good separation characteristics and thus may provide an improved membrane with enhanced qualities.

FTIR and other analyses[39] confirmed the existence of intermolecular hydrogen bonding between PBI and PA; the carbonyl band shifts from 1741 to 1730 cm^{-1} in an 80/20 blend. Also, miscibility and compatibility were indicated further by the clarity (visual and microscopic examination) of films prepared from the 80/20 blend; TGAs showed blends conforming to the Rule of Mixtures, and X-ray diffraction (XRD) analysis confirmed that PBI and PA interact and are compatible on a molecular scale; blends were found to have physical properties adhering to the Rule of Mixtures values; the thermal stability of the blends at 550°C increased proportionately as the relative PBI concentration increased; and after blending, the hydrophilicity of the PBI decreased dramatically. Thus, PAs and aromatic PBIs were found to be miscible in all proportions, and synergistic effects of one polymer over the other's weak properties can be achieved.

PA is soluble in common organic solvents such as methylene chloride and tetrahydrofuran, whereas PBI is insoluble in these solvents. The two polymers can be simultaneously dissolved in a common solvent (DMF, DMAc, DMSO, NMP, concentrated acid, such as sulfuric), or solution-derived miscible compositions of each polymer may be separately dissolved in separate volumes of a mutual solvent, and desired proportions of the resulting solutions can be mixed together to form a solution of the two polymers. If the blend is mostly PBI, the less expensive polyarylate should be added to an extent to take advantage of rendering the blend more tractable and processable, and less susceptible to moisture. If the blend is mostly polyarylate, then one should take advantage of PBI's ability to render the blend less sensitive to solvents, particularly chlorinated hydrocarbons. The preferred blend, according to the referenced patent, falls somewhere around 40 to 60 parts by weight of PBI. Articles in the form of films, fibers, fibrets, and molded structures may be prepared from such blends. Molding compositions may also incorporate into the polymer matrix ~1 to 50 wt%, preferably 10 to 30 wt% based on the total weight of the molding composition, of a solid filler and/or reinforcing agent, i.e. glass fibers, carbon fibers, synthetic polymer fibers, aluminum fibers, wool fibers, silica, clays, mica, talc, graphite, etc.[38,40]

7.8 PBI–polysulfone (PS) blends

PBI and polysulfone (PS) do not form miscible blends,[41] but the tensile modulus and strength of their blend fibers were found to be comparable to, or better than, that of PBI fibers, depending on the process conditions. Fiber spun from 80/20 PBI–PS solution blends were found to have PBI-like limiting oxygen indices (LOI) and flame shrinkage behavior. The 80/20 blend fibers were found to have comparable or superior mechanical performance to that of miscible PBI–Ultem blend fibers, and with higher modulus but lower elongation than stabilized PBI.

Chung and others also reported[42] stable solutions of PBI (70–95 wt%) and PS (5–30 wt%) blends, substantially free of the typical salt stabilizers, such solutions being useful for the production of films, fibers, and molded articles exhibiting enhanced mechanical properties. Polysulfones are useful tractable polymers, but the degree of reduction in thermal and mechanical properties is usually proportional to the wt% of the melt-processable polymer addition. Blends of PBI and PS are generally prepared by solution blending to form a dope solution for extrusion or casting; alternatively, the resin blend may be precipitated, recovered, dried, and molded, as is or after milling. It was determined that small amounts of PS to PBI can decrease moisture takeup (typically at ~15%), reduce the processing temperature, increase the tractability, and reduce the cost of resulting blend products.

It has been reported[43] that the introduction of functional groups in the polymer chain of PS, such as sulfonate groups, can result in the formation of miscible blends with PBI. Thus, the miscibility behavior of a series of blends of PBI and sulfonated PS at various sulfonation levels has been studied by dynamic mechanic analysis (DMA), FTIR, and FT-Raman spectroscopy. Partially miscible or miscible blends were obtained when the sulfonation level was higher than 10 mol%. FTIR band shifts associated with the N-H and sulfonate groups were shown to account for the induced miscibility in terms of specific interactions. Also, preliminary results were shown to demonstrate the potential usefulness of PBI–PS polymer membranes as electrolytes for fuel cells.

Sulfonation of PS has been found to cause moderate increases in the T_g from 190°C to 210–225°C, and with the PBI–sulfonated PS miscible blend exhibiting excellent miscibility characteristics based on the hydrogen bonding between the N-H group of PBI and the sulfonated groups of the PS. Thermal stabilities at over 200°C were achievable, and the miscibility was found to be dependent on the PBI–sulfonated PS ratio as well as the degree of sulfonation.[44,45]

7.9 PBI–polyimide (PI) and PBI–polyamide-imide (PAI) blends

PBI may be blended with polyimides (PIs) to improve processability, and thermoplastic PIs can offer improved thermo-oxidative stability. However, PIs often lack the thermal and mechanical attributes required to perform in high use-temperature applications (>300°C). Thus, blending with PBI can accomplish intermediate thermal and mechanical properties.

7.9.1 PBI–PI blends

Various miscible blends based on aromatic PBI and commercially available PIs have been reported.[7, 46,48] Blend miscibility was evidenced by infrared (IR) spectra and a single T_g lying between that of the constituent polymers. The intermolecular interactions involving the N-H (from PBI) and carbonyl groups (from PI) were the major driving forces for the miscibility affording the possibility for enhanced melt processing of such blends.[49] The IR spectral shifts are eliminated upon phase separation.[50,51] Previous efforts[52] found PBI to be miscible with synthetic PI blends of poly (amic acids) using DMAc co-solvent, cast film or precipitated into powder. Single composition dependent T_g values between those of the homopolymers were determined, and miscibility was determined to arise solely from hydrogen bonding between the carbonyl from the 5-membered ring and N-H.

Chenevey and others[53] reported on homogeneous blends of PBIs and PIs, obtained by mixing in a mutual solvent, for conversion into solvent-resistant films and fibers having good properties. The blends comprised 25–40 wt% PBI (intrinsic viscosity 0.7 to 1.2 in 97% sulfuric acid) and 75 to 60 wt% PI, soluble in DMAc, DMF, and DMSO and, optionally, a stabilizer such as lithium chloride and other agents may be added to the blend solution to promote stability during mixing and storage. Alternatively, such blends may be prepared by melt mixing solid dry mixtures of PBI and PI in devices such as heated screw mixers that provide efficient mixing and heating with limited residence times such as to minimize degradation at high temperature. The blends were reported to be highly solvent resistant, highly resistant to flame embrittlement, and to have single T_g values (at least 270°C). Additionally, the blends were found to retain the desirable features of the PI, including high heat resistance, high strength-to-weight ratio and modulus, and excellent non-flammability characteristics and processability. The blends were found suitable for formation of fibers, films, fibrids, and compression molded articles. Typical fiber properties were reported at or over 3.5 g/denier at 25°C, elongation of at least 25%, and modulus of at least 65 g/denier. Film properties included tensile strength of at least 12 000 psi, elongation of at least 40%, and modulus of at least 300 000 psi.

PBI and PI have been found to be soluble in DMAc, as well as DMF and NMP, and thus blends can be prepared by solution mixing, followed either by precipitation into a non-solvent like methanol or by casting a film. The T_g of a 50/50 blend has been measured at ~380°C. The ratio of PI in the blend should be governed by the level of moldability needed and the decreased susceptibility to moisture regain required of the application. Heat-induced cross-linking may lead to further utility of such blends, such as modified softening temperature and solvent resistance under stress-enhanced T_g with aging conditions.[54]

7.9.2 PBI–PAI blends

Polyamide-imides (PAIs) are thermoplastic amorphous polymers prepared by the condensation of an aromatic diamine, such as methylene diamine, and an anhydride, such as trimellitic acid chloride. PAIs have good mechanical, thermal, chemical resistance, high strength, melt processability, and high heat capacity. They can be processed into a variety of forms, such as injection or compression molded articles, coatings, films, fiber, and adhesives. The typical heat deflection temperature for neat molded PAI is ~278°C, but reinforcements are often used to improve mechanical properties. PAIs are generally soluble in strong aprotic solvents such as NMP and DMAc, and thus miscible blends with PBI are feasible.

Chen and others[55] reported on blends of PBIs and polyamide-imides having fluorine-containing linking groups, 5–95 wt% PBI. The F-containing linking groups were prepared by forming the polycondensation product of one or more aromatic diamines, one or more trifunctional acid anhydride monomers, and one or more tetrafunctional aromatic dianhydrides, at least one of the monomers containing the group, CF_3-$C(H_2)$-R, linking two aromatic moieties where R is CF_3 or phenyl. PAI typically has a T_g of ~275°C;

however, it exhibits poor flow properties, rendering difficulties with injection molding and spinning into fibers. The fluorine-containing linking group improves flow characteristics, improves solubility in most organic solvents, such a trichloroethylene as compared to PI, and improves hydrophobic properties as well as thermo-oxidative stability. Thus, applications in wire enamel coatings, laminates, molded products, films, fibers and impregnating varnishes may be significantly improved, including thermal stability, particularly when blended with PBI. Such blends with fillers, reinforcing agents, and post-heat treatment with or without sulfuric acid, can be used in film and fiber forms where shrinkage can be minimized and resistance to solvents and acids is enhanced.

7.10 PBI–poly (bisphenol-A carbonate) (PC) and PBI–polybenzoxazole (PBO) blends

Polycarbonate blends with PBI may be used to improve the blend miscibility by intermolecular interactions (hydrogen bonding) as compared to polyimides, and subsequently improved relative thermal stability. Due to the high temperature performance attributes of PBI and PBO, blending would be expected to enhance further the thermal characteristics and end-uses, and with good solubility in typical solvents for the individual polymer resins.

7.10.1 PBI–PC blends

Musto and others[56] pursued PBI blended with poly (bisphenol-A carbonate) (PC), poly-[2,2-propane-bis(4-phenyl carbonate)], as a means to investigate polymers that are comparatively stronger proton acceptors compared to PI. PBI possesses a strong tendency to form hydrogen bonds acting as proton donors, and thus bisphenol A was selected as a candidate toward enhancing the strength of the intermolecular interactions with PBI and thus improving miscibility, in addition to its reactivity, thermal stability, and ability to form miscible blends. Hydrogen bonding interactions occur among the N-H groups of PBI and the carbonyl groups of PC as evidenced from FTIR spectra. The T_g of the blend was found to occur at ~240°C as opposed to ~220°C calculated from the Fox equation. Thus, PBI was found to have a profound impact on the stability of PC, and with PC degradation in the presence of PBI occurring mainly by dissociation of the carbonate groups via catalyzed hydrolysis. Based on an FTIR quantitative analysis of the carbonyl region of a 50/50 wt% blend, the estimated T_g of the blend was found to be close to the value predicted by the Fox equation.

7.10.2 PBI–PBO blends

Attempts have been made to combine PBI and polybenzoxazole (PBO) to obtain more thermally stable and tractable polymers.[57] Both polymers have received attention due to their chemical resistance, thermal stability, and good mechanical properties at high temperature. Co-polymers have also been reported,[58,59] with the claim that phase separation would not easily take place compared with solution mixing/blending of the two polymers. It was found that the initial decomposition temperature of the co-polymer was ~590°C, with 5% weight loss at 590–600°C, and with ~50 wt% of residue remaining at 770°C, demonstrating reasonably good thermal stability. The co-polymer was found to be soluble in hot DMF, DMAc, NMP, and sulfuric acid.

7.11 PBI–poly(4-vinyl pyridine) (PVPy) and other blends

The good miscibility of PBI–PVPy blends can be useful for property modifications in the plastics industry, particularly in the areas of processability, where intermediate T_g values are sufficient and economical, and with retention of good mechanical properties at elevated temperatures. Blending can be achieved by solution mixing, and hydrogen bonding promotes compatibility.

7.11.1 PBI–poly(4-vinyl pyridine) (PVPy) blends

In pursuit of high temperature polymer blends, the miscibility of such polymer systems have been studied previously.[7,47,48,60] PVPy was further investigated,[61] based on pyridine being a strong base capable of forming hydrogen bonds with the N-H group of PBI. Blends were prepared by solution mixing in DMAc, and PBI was shown to be miscible with PVPy over the entire composition range. Blends of all compositions were shown to have a single T_g, lying close to or above the weight average T_g of the two components. PVPy was found to begin to decompose at temperatures over 375°C. FTIR showed that miscibility was promoted by interaction of H-bonding, and even small amounts of H-bonding may be sufficient to promote miscibility. Such blending may have utility in the pursuit of useful property modifications.

7.11.2 Other blends

Calundann and others[62] reported on novel miscible blends comprising from about 5 to 95 wt% of PBI and 95 to 5 wt% of an aromatic polyamide, aromatic polyamide-hydrazide or aromatic polyamide containing heterocyclic linkages, preferably at least 50 to about 95 wt% of the PBI and from 50 to 5 wt% of the polyamide component. Useful blend solvents include NMP, DMF, DMAc, and DMSO; heating may be required to obtain clear viscous solutions. The polyamide-type components were characterized as having high T_g values and HDTs, but with inferior non-flammability characteristics. The non-PBI component was recommended to be present in the blend, mostly PBI, at a sufficient level to result in a blend with improved physical properties (tensile strength and modulus) relative to PBI, more thermally tractable than PBI, and also less susceptible to moisture, and while maintaining the attractive features of PBI. The PBI level in a blend consisting mostly of the polyamide component was recommended at a level such that the blend had increased thermal stability and non-flammability relative to the polyamide component. Such blends were found to have good thermal, flame, and solvent resistance, improved physical properties, good price/performance characteristics, and to be useful in severe environmental conditions. Moldable blends may be prepared by freeze milling the dry blends to 100 microns or less in diameter. Blends in the form of films, fibers, fibrets, or molded articles may be post-treated with heat or sulfonating agents in order to minimize shrinkage and increase solvent and acid resistance.

7.12 PBI commercial products

For engineers and researchers searching for materials for specific high temperature applications, data and information about commercially available PBI and PBI blends are important. PBI commercial products include PBI fibers, PBI polymers, compression molded PBI products, melt-processable PBI/PEEK blends, and PBI solutions. Their typical properties are summarized in this section.

7.12.1 Commercial PBI fibers

The commercially available PBI fibers are 1.5 denier staple and short cut fibers. The typical properties of PBI fibers are shown in Table 7.1. PBI fibers are highly thermally stable and maintain good strength after exposure to flame. They do not burn, melt or drip. The natural color of PBI fiber is golden brown. Black PBI fibers have been developed recently. PBI fibers can be easily blended with other staple fibers during yarn spinning or nonwoven manufacturing. Para-aramid and FR-Rayon fibers are used most to blend with PBI fibers to produce high performance flame resistant fabrics. Since the 1990s, more and more cities have chosen the fabrics made of 40% PBI with 60% para-aramid as the outer shells for fire fighter garments because

Table 7.1 Properties of PBI staple fibers

Property	Imperial unit	Metric unit
Denier per filament (DPF)	1.5 denier	1.7 dtex
Tenacity	2.7 g/d	2.4 dN/tex
Breaking elongation	27%	27%
Initial modulus	45 g/d	40 dN/tex
Crimp	9.5/inch	3.7/cm
Specific gravity	1.4	1.4
Moisture regain at 68°F (20°C) 65% RH	15%	15%
Boiling-water shrinkage	<1.0%	<1.0%
Hot-air shrinkage at 400°F (205°C)	<1.0%	<1.0%
Specific heat	0.3 BTU/lb F	1.3 kJ/(kg K)
Limiting oxygen index	>41%	>41%
Standard cut length*	1/8, ¼, 1½, 2, 3, 4 inch	3, 6, 38, 50, 76, 102 mm

*Other cut lengths may also be available.

they outperform other fabrics in fire protection while providing more comfort to wear. Sold under the trade names of PBI Gold® and PBI Matrix®, these PBI fabrics have reduced burn injuries and saved many lives.

7.12.2 Commercial PBI polymers

Commercial PBI polymers are in powder form and are golden in color. A moderately low molecular weight fraction PBI is generally used for molding. Higher molecular weight fractions are utilized for solutions, films, coatings and membranes.

7.12.3 Compression molded PBI products

Compression molded PBI stock shapes are commercially available under the trade name Celazole® U-60. High temperature and high pressure are needed to compression mold PBI. PBI stock shapes can be further machined into specific parts. With the highest T_g among thermoplastics at about 427°C (DMA method), Celazole® U-60 also has the highest compressive strength at 57 kpsi. The superior physical properties, high temperature properties and outstanding chemical resistance make Celazole® U-60 the world's highest performing plastic material. Typical properties data of Celazole® U-60 are summarized in Table 7.2.

Celazole® U-60SD is a special grade of PBI products with extremely low metallic ion contents (<1 ppm per species) and is generally used in the

Table 7.2 PBI Celazole® U-60 typical properties

Properties	ASTM method	Imperial value	Metric value
Mechanical			
Tensile strength	D-638	23 kpsi	160 MPa
Modulus		850 kpsi	5900 MPa
Elongation		3.0%	3.0%
Tensile fatigue, % of stress to failure at			
1 000 000 cycles, 1 Hz	D-638	35% (8.1 kpsi)	35% (56 MPa)
Flexural strength	D-790	32 kpsi	220 MPa
Modulus		950 kpsi	6500 MPa
Compressive strength (yield)	D-695	57 kpsi	390 MPa
Compressive strength (10% strain)	D-695	50 kpsi	340 MPa
Compressive modulus	D-695	850 kpsi	5900 MPa
Hardness – Rockwell M	D-785	>125	>125
– Rockwell E	D-785	104	104
– Shore D	D-2240	95	95
Izod impact strength (notched)	D-256	0.53 ft-lb/in	30 J/m
(unnotched)		11 ft-lb/in	590J/m
Thermal			
Heat deflection temp. (264 psi; 1.8 MPa)	D-648	815°F	435°C
Glass transition temperature	DMA	800°F	427°C
Coefficient of linear thermal expansion			
75–300°F (25–150°C)	TMA	13×10^{-6} in/in°F	23 μm/m °C
390–570°F (200–300°C)	TMA	18×10^{-6} in/in°F	33 μm/m °C
Limiting oxygen index	D-2863	58%	58%
Thermal conductivity 77°F (25°C)		2.8 Btu in/hr-ft^2 °F	0.41 W/m °C
Electrical			
Dielectric strength	D-149	580 V/mil	23 kV/mm
Volume resistivity	D-257	2×10^{15} ohm-cm	2×10^{15} ohm-cm
Dissipation factor			
1 kHz	D-150	0.000	0.000
10 kHz	D-150	0.003	0.003
0.1 MHz	D-150	0.034	0.034
Dielectric constant			
1 kHz	D-150	3.4	3.4
10 kHz	D-150	3.4	3.4
0.1 MHz	D-150	3.3	3.3
Arc resistance	D-495	185 s	185 s
Other			
Specific gravity		1.3	1.3

(*Continued*)

Table 7.2 Continued

Properties	ASTM method	Imperial value	Metric value
Coeff. of friction, static			
Aluminum		0.14	0.14
Steel		0.15	0.15
Brass		0.16	0.16
Coeff. of friction, dynamic			
Aluminum		0.16	0.16
Steel		0.16	0.16
Brass		0.18	0.18
Water adsorption, 24 h at 73°F	D-570	0.4%	0.4%

semiconductor industry due to its high purity. Its mechanical properties are similar to U-60. Other U-60 grades include U-60ESD, an electrostatic dissipative grade used to prevent static buildup in the manufacture of certain electronic parts, and U-60CF, a high strength, high modulus form of U-60. Moisture absorption of PBI is higher than many other thermoplastics. PBI parts should be dried before use in applications in which parts are heated rapidly to above 300°C.

7.12.4 Commercial PBI–polyetheretherketone (PEEK) blends

As discussed in Section 7.2, PBI polymers can be compounded with many other high temperature plastics to make different PBI blends. Currently, four PBI–PEEK(polyetheretherketone) blends are commercially available under the following trade names:

- Celazole® TU-60
- Celazole® TL-60
- Celazole® TF-60C
- Celazole® TF-60V.

These four blends are sold in pellet form for injection-molding applications. Typical properties of molded Celazole® TU-60, TL-60, TF-60C, and TF-60V products are listed in Table 7.3.

Celazole® TU-60 is an unfilled blend of PBI–PEEK. Independently, PEEK is a high performance thermoplastic with excellent mechanical properties, chemical resistance, and thermal stability. The TU-60 blend has even better thermo-mechanical properties than PEEK, particularly at elevated temperatures. And TU-60 can be injection molded to make parts at lower costs than PBI.

Celazole® TL-60 is a self-lubricating grade of the PBI–PEEK blend. It is an injection-molding grade like TU-60 but with added features of low

Table 7.3 Typical properties for commercial PBI–PEEK blends

Properties	ASTM test method	TU-60 Unreinforced Melt processable		TL-60 Self-lubricating Melt processable		TF-60C Carbon fiber reinforced Melt processable		TF-60V Glass fiber reinforced Melt processable	
		Imperial value	Metric value	Imperial value	Metric value	Imperial value	Metric value	Imperial value	Metric value
Tensile strength	D-638	14.5 kpsi	100 MPa	16 kpsi	110 MPa	33 kpsi	230 MPa	22 kpsi	150 MPa
Modulus		730 kpsi	5000 MPa	2600 kpsi	18,000 MPa	3500 kpsi	24,000 MPa	1750 kpsi	12000 MPa
Elongation		2.2%	2.2%	0.9%	0.9%	1.4%	1.4%	1.4%	1.4%
Flexural strength	D-790	25.4 kpsi	175 MPa	23 kpsi	158 MPa	46.5 kpsi	320 MPa	32.6 kpsi	225 MPa
Modulus		725 kpsi	5000 MPa	1900 kpsi	13,000 MPa	3050 kpsi	21000 MPa	1810 kpsi	12500 MPa
Compressive strength (yield)	D-695	30 kpsi	206 MPa	18 kpsi	123 MPa	32 kpsi	220 MPa	32 kpsi	220 MPa
Modulus		430 kpsi	2900 MPa	740 kpsi	5100 MPa	550 kpsi	3800 MPa	530 kpsi	3600 MPa
Hardness – Rockwell A	D-785	25	25	24	24	30	30	30	30
Specific gravity		1.3	1.3	1.43	1.43	1.41	1.41	1.52	1.52

(*Continued*)

Table 7.3 Continued

Properties	ASTM test method	TU-60 Unreinforced Melt processable		TL-60 Self-lubricating Melt processable		TF-60C Carbon fiber reinforced Melt processable		TF-60V Glass fiber reinforced Melt processable	
		Imperial value	Metric value	Imperial value	Metric value	Imperial value	Metric value	Imperial value	Metric value
Poisson's ratio				0.42	0.42	0.39	0.39	0.34	0.34
Coefficient of thermal expansion 75–300°F	TMA	19×10^{-6} in/in °F	34 μm/m °C	14×10^{-6} in/in °F	26 um/m °C	14×10^{-6}in/in °F	26 um/m °C	9×10^{-6} in/in °F	17 um/m °C
Temperature of initial (5%) weight loss in air	TGA	930°F	499°C	1116°F	602°C	1110°F	599°C	1125°F	607°C
– in nitrogen	TGA	930°F	499°C	1123°F	606°C	1160°F	627°C	1175°F	635°C
Thermal conductivity	F-433			5.3 Btu in/h ft^2 °F	0.76 W/m °C	3.1 Btu in/h ft^2 °F	0.45 W/m °C	2.5 Btu in/h ft^2 °F	0.36 W/m °C
Heat capacity at 158°F	DSC			0.28 Btu/lb °F	1160 J/kg °C	0.28 Btu/lb °F	1180 J/kg °C	0.27 Btu/lb °F	1130 J/kg °C
Heat deflection temperature at 264 psi	D-648	482°F	250°C	590°F	310°C	608°F	320°C	590°F	310°C

friction and high wear resistance. The ideal applications are for bearings, bushings, and seals where friction, heat and chemicals can shorten the life of other plastic materials.

Celazole® TF-60C is a carbon-reinforced PBI/PEEK blend. It is an injection-molding grade like TU-60, but with much higher modulus and strength. The improved physical properties are particularly suitable for applications that require superior strength and low deformation at high temperature. It is also static dissipative.

Celazole® TF-60V is a glass-reinforced PBI/PEEK blend. It is an injection-molding grade with properties similar to TF-60C except that TF-60V is a much better thermal and electrical insulator. It is ideally suited for applications that require high strength, lower deformation, and excellent insulation under high temperatures.

7.12.5 PBI solutions

Homogeneous PBI solutions in DMAc (dimethylacetamide) solvent are commercially available at 26% and 10% concentrations. PBI solutions are used for fiber spinning, coating, impregnation, film casting and prepreg applications. The 26% PBI/DMAc solution contains LiCl to keep the solution stable. Solution viscosities at different temperatures are shown in Fig. 7.13. The 26% PBI/DMAc solution can be easily diluted to a lower concentration by mixing with additional DMAc solvent. PBI solution viscosities at different concentrations are shown in Table 7.4. The 10% PBI solution is made without LiCl and is easier to be applied during coating processes.

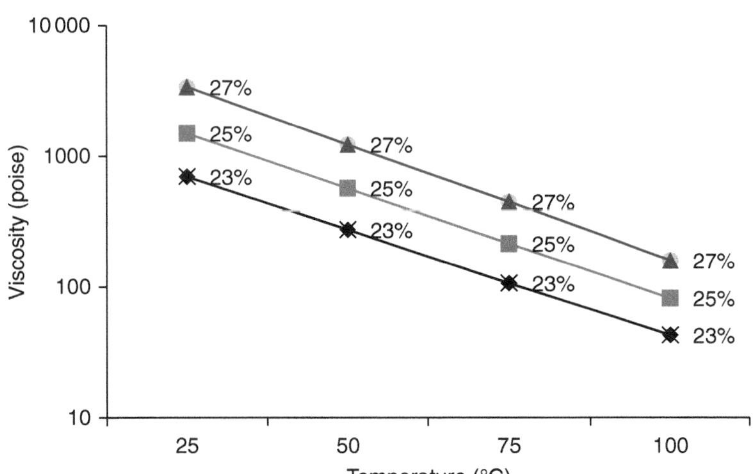

7.13 PBI/DMAc solution viscosity at different temperatures.

Table 7.4 Viscosities of PBI/DMAc solutions at different concentrations (25°C)

PBI solids (%)	Viscosity (poise)
25	2000
20	175
15	20
10	2
5	0.2

7.13 PBI in high temperature applications

PBI finds many high temperature applications, not only in protecting fire fighters and military personal, but also in semiconductor, clean energy, oil, gas, chemical, and aerospace industries because of its extremely thermally stable molecular structure and superior physical properties. Examples of some of the applications are briefly discussed in this section.

7.13.1 Molded products of PBI and PBI blends

Compression-molded PBI shapes can be machined into parts of various complexities. As the world's highest performing thermoplastics, PBI parts are often used in the most demanding environments where other polymer materials fail to meet the temperature and performance requirements. For example, in the dry etch chambers for semiconductor wafer manufacturing, PBI parts are selected because of their high temperature dimensional stability, wear resistance, and plasma resistance. PBI seals, lifting pins, bushing, screws, insulators, and other components have replaced parts made of Quartz, Alumina and polyimide in the plasma etch chambers, providing prolonged life cycle and reduced wafer contamination. PBI parts are also used in flat glass panel manufacturing processes as clamps, guides, and other components. The longer life cycle, superior dimensional stability at high temperature, and the ability to control electrostatic discharge make PBI parts excellent choices.

PBI–PEEK blends TU-60, TL-60, TF-60C and TF-60V are often used for applications where materials with performance superior to PEEK are required. The melt-processable PBI–PEEK blends are cost competitive options to meet the challenges of those applications. Celazole® TL-60, a PBI–PEEK blend with enhanced lubrication properties, works especially well for parts with rotating or sliding surfaces such as bearings and bushings. A study[63] examining six high performance plastics, including PBI TL-60, PEEK, polyimide (PI), and polyamide-imide (PAI) through high PV wear tests demonstrated this. The results showed that the TL-60 sample had the lowest coefficients of friction and lowest wear factor. It was the only material in the tests that survived loads at or above 2500 psi at 50 ft/min speed.[63]

Examples of PBI–PEEK products can be found in bearings, bushings, gears, piston rings, washers, seals, pads, rollers, conveyors, and many other molded parts.

7.13.2 Fabrics made of PBI fiber blends

The fabrics made of 40% PBI with 60% Para-aramid fiber blend are the premier outershell materials for fire fighter suits. PBI fibers improve a fabric's strength after flame exposure, increase comfort for users, and reduce heat transfer through the fabric. As a result, the PBI–para-aramid fabrics are the first choice for professional firefighters in most large cities in the US, Canada and many cities in Europe. There is an increasing awareness of PBI fabrics in Asia Pacific. Hong Kong, South Korea, Saudi Arabia, and New Zealand are examples of the places in Asia using PBI fabrics to protect their firefighters.

Protection from fire, arc flash, explosions, and other hazards is very important for military personal and for workers in the electrical, oil, gas, and chemical industries. Fabrics made of PBI–para-aramid–FR-Rayon fiber blends perform very well in these applications. Also, PBI–FR-Rayon fiber blends can be knitted to make a soft fabric ideal for protective undergarments or hoods which provide both protection and comfort.

7.13.3 Examples of other applications

PBI short cut fibers have been used to replace asbestos in high temperature applications such as rocket motor insulation. Rubber formulas incorporating PBI short cut fibers showed superior performance in a development of ablative materials for NASA solid fuel rocket motor insulation. As a result, PBI short cut fibers are specified for ablative insulation in the development of new solid fuel rocket motors. Nonwoven fabrics containing PBI are also excellent materials for the fire blocking layers in airplane seats due to their better performance at lighter weight.

PBI membranes have been the critical components in high temperature polymeric electrolyte membrane fuel cells. PBI fuel cells tolerate much more impurities in the H_2 fuel compared to low temperature fuel cells. Therefore, they are preferred in fuel cell systems where the H_2 supplies are from reformers converting other fuels to H_2. PBI fuel cells are also suitable for combined heat and power fuel cell systems. In other applications, PBI membranes have been successfully demonstrated to purify H_2 while pumping it against a high pressure. The technology can be applied to recover H_2 in metal heat treating ovens as well as many other processes that use H_2.

In liquid separation, hollow fiber membranes based on PBI have shown excellent performance for pervaporation dehydration of organic liquids.[64–67] For example, a dual layer PEI–PBI hollow fiber membrane with an outer selective layer of PBI showed better performance than most other polymeric membranes in pervaporation dehydration of ethylene glycol.[67] Sulfonation modifications of PBI membranes have demonstrated excellent separation efficacies in the dehydration of acetic acid.[68] Studies have shown that PBI hollow fiber membranes were effective in separating chromates from solutions.[69] Also, PBI nanofiltration hollow fiber membranes are promising candidates as forward osmosis membranes.[70] In gas separation, recent studies sponsored by the Department of Energy at Los Alamos National Laboratories and SRI International demonstrated potential applications of PBI membranes in carbon capture and H_2 purification from synthesis gas streams at elevated temperatures.[71] H_2/CO_2 selectivity > 40 has been achieved at H_2 permeability of 200 GPU at 250°C.[72]

7.13.4 PBI composites and coatings

With the highest T_g among commercial thermoplastics, PBI can be used as the matrix resin to make composite products with superior thermal properties. The potential exists to increase the temperature range for advanced polymer composites through further development of PBI composite materials.

PBI coatings are highly resistant to heat, chemicals and wear. They are suitable for protecting metal surfaces working under friction at high temperature. US patent 5549946 describes how the surface of a heat-resistant roll for a copying machine is coated with a varnish containing polybenzimidazole as a major resin component.[73] US patent 567414 describes how an electrical wire is coated with a polybenzimidazole varnish.[74] US Patent Application 12/957601 describes a mixture of polybenzimidazole and polyacrylate.[75] US patent 8475925 describes a coating based on polybenzimidazole with superior adhesion characteristics through the inclusion of an epoxy component.[76] All of these patents/patent applications advance PBI as a coating by increasing the versatility as a conformal or planar coating in stressful thermal mechanical systems.

7.14 Future trends

PBI's superior properties, such as high glass transition temperature (T_g), good chemical resistance and excellent mechanical properties at both high and cryogenic temperatures, allow products made of PBI to continue outperforming those made of other commercial high temperature polymers. One trend for PBI application is replacing other materials to improve the performance of an existing high temperature product. In the meantime, technology advancements in defense, aerospace, energy, semiconductor and other industries are demanding material properties that many other performance polymers cannot provide. PBI products are good candidates to meet these new challenges.

Applications of the melt-processable commercial PBI/PEEK blends TU-60, TL-60, TF-60C, and TF-60V are growing. Improvements in processing technologies for these PBI blends are leading to more products with higher quality and efficiency. Knowledge of PBI materials with understanding of specific applications can lead to developments maximizing the benefits of PBI blends for different end uses. Moreover, new and unique properties can be created within targeted performance and price specifications by well-designed PBI blends. Developments of new commercial PBI blends in the future will provide more choices for challenging high temperature applications.

In the protective clothing arena, the trend is continuing for firefighters in more countries to select PBI fabrics for their protective gear. For over two decades, PBI fabrics have been established in North America as the best choice for firefighter turnouts due to their superior overall performance. In recent years, the use of PBI fabrics in the fire services is growing fastest outside North America as countries around the world are seeking better protection for their firefighters. Additionally, there are continuing developments of different PBI fiber blends and fabrics for uses beyond the fire service market. The properties, comfort, and protective performances of these fabrics make them suitable for protective garments for military and police personnel as well as for workers in chemical, oil, gas, and electric industries. An ultra-high performance, unique polymer, PBI will make more and more contributions in high temperature applications as innovations continue in PBI materials, process engineering, and end-use applications.

7.15 Sources of further information and advice

See: www.pbiproducts.com.

7.16 References

1. H. Vogel and C.S. Marvel, *J. Polym. Sci.* **50**, 511 (1961).
2. C. S. Marvel and H. Vogel, US Patent 3, 174,947 (1965); *Chem. Abstr.*, **63**, 7137 (1965).
3. Celanese internal communications.

4. B.L. Oslin, US Patent 4,933,140 (1990).
5. E. W. Choe and D. D. Choe, 'Polybenzimidazole (Overview)' in *Polymeric Materials Encyclopedia*, vol 7, J.C. Salamone, Ed., CRC Press, 5619 (1996).
6. M. Jaffe, P. Chen, E.W. Choe, T.S. Chung, and S. Makhija, 'High Performance Polymer Blends', *Adv. Polym. Sci.*, **117**, 297 (1994).
7. L. Leung, D. J. Williams. F. E. Karasz, and W.J. MacKnight, *Polymer Bull.*, **16**, 457 (1986).
8. T. S. Chung, 'Polybenzimidazoles', in *Handbook of Thermoplastics*, O. Olabisi, Ed., Marcel Dekker Inc., (1997).
9. E. J. Powers and G. A. Serad, *History and Development of Polybenzimidazoles, High Performance Polymers: Their Origin and Development*, R. B. Seymour and G. S. Kirshenbaum, Eds., Elsevier, 355, (1986).
10. E.W. Choe, 'Catalysts for the Preparation of Polybenzimidazoles', *J. Appli. Polym. Sci.*, **53**, 497 (1994).
11. E.W. Neuse, 'Aromatic Polybenzimidazoles; Synthesis, Properties, and Applications', *Adv. Poly. Sci.*, **47**, 1, (1982).
12. H.H. Levine, *Encycl. Polymer Sci. Technol.*, **11**, (1969).
13. J.F. Jones, *Macromol. Sci.*, **C2**, 303 (1968).
14. P. E. Cassady, *Thermally Stable Polymers, Synthesis and Properties*, Marcel Dekker Inc., New York, 168, (1980).
15. B. C. Ward, *SAMPE Journal*, **25**(2), Mar/April (1989).
16. Y.F. Wang, T. Hsu, A.S. Hay, K. Li., and B. Patel, US Patent Application 2011/01014417(2011).
17. T.S. Chung, W.F. Guo, and Y. Liu, *J. Membrane Sci.*, **271**, 221 (2006).
18. T.S. Chung and Z.L. Xu, *J. Membrane Sci.*, **147**(1), 35 (1998).
19. S.S. Sudhir and D.W. Kratzer, US Patent 7,950,529 (2011).
20. P.J. Flory, *Principles of Polymer Chemistry*, Cornell University Press, Chapter 13, 555 (1953).
21. F. W. Billmeyer, Jr., *Textbook of Polymer Science*, Wiley-Interscience, 176 (1984).
22. P.N. Chen, T.S. Chung, P.J. Harget, and M.J. Jaffe, *Polym. Mater. Sci. Eng.*, **59**, 707 (1988).
23. P. Chen and R. Tucker, US Patent 4,628,067 (1986).
24. B. G. Dawkins, M Gruender, G.S. Copeland, and T. Hsu, US Patent 7,915,351 (2011).
25. P. Musto, F.E. Karasz, and W. J. MacKnight, *Macromolecules*, **24**, 4762 (1991).
26. B. de Jong, A.J. Waddon, F.E. Karasz, and W. J. MacKnight, *Polym. Eng. Sci.*, **32** (15), 1047 (1992).
27. D. L. VanderHart, G.C. Campbell, and R.M. Briber, *Macromolecules*, **25**, 4734 (1992).
28. L. Leung, D. J. Williams. F.E. Farasz, M. Jaffe, and W.J. MacKnight, US Patent 4,973,630 (1990).
29. D.M. White, T. Takekoshi, F. J. Williams, H.M. Relles, P.E. Donahue, H.J. Klopter, G. R. Loucks, J.S. Manello, R.O. Matthewa, and R.W. Schluenz, *J. Polymer Sci., Polym. Schem. Ed.*, **19**, 1635 (1981).
30. B. G. Dawkins, M. Gruender, G.S. Copeland, and J. Zucker, US Patent 7,629,420 (2009).
31. E. Alverez, L.P. DiSano, and B.C. Ward, US Patent 4,912,176 (1990).
32. T.S. Chung and. P.N. Chen, *Polym. Eng. Sci.*, **30**, 1 (1990).

33. T.S. Chung and P.N. Chen, *J. Appli. Polym. Sci.*, **40**, 1209 (1990).
34. T.S. Chung and F.K. Herold, *Polym. Eng. Sci.*, **31**, 1520 (1991).
35. L.M. Robeson, US Patent 4,259,458 (1981).
36. L. M. Robeson, M. Matzner, and L.M. Maresca, US Patent 4,250,279 (1981).
37. L.M. Robeson, W.D. Claus, Jr., and H.L. Batleman, US Patent 4,369,136.
38. T.S. Chung, P.N. Chen, and V.J. Provino, US Patent 4,842,740 (1989).
39. P.N. Chen, T.S. Chung, and P.J. Harget, US Patent 4,963,628 (1990).
40. P.N. Chen, P.J. Harget, and T.J. Dolce, US Patent 4,963,434 (1990).
41. T.S. Chung, M. Glick, and E.J. Powers, *Polym. Eng. And Sci.*, **33** (16) 1042 (1993).
42. T.S. Chung, A.G. Schlask, and D. Kurschus, US Patent 5,208,298 (1993).
43. V. Deimede, G.A. Voyiatzis, J.K. Kallitsis, L. Qingfern, N.J. Bjerrum, *Macromolecules*, **33** (20), 7609 (2000).
44. C. Hasiotis, V. Deimede, and C. Kontoyannis, *Electrochemica Acta*, **46**, 2401 (2001).
45. B. C. Johnson, I. Yilgor, C. Tranc, M. Iqbal, J.P. Wright, D.R. Lloyd, and J.E. McGrath, *J. Polym. Sci. Polym. Chem. Ed.*, **22**, 721 (1984).
46. S. Stankovic, G. Guerra, D.J. Williams, F.E. Karasz, and W.J. MacKnight, *Polym. Comm.*, **29**, 14 (1988).
47. G. Guerra, D.J. Williams, F.E Karasz, and W.J MacKnight, *J. Polym. Sci, Phys. Ed.*, **26**, 301 (1988).
48. G. Guerra, S. Choe, D.J. Williams, F.E Karasz, and W.J. MacKnight, *Macromolecules*, **21**, 231 (1988).
49. W.J. MacKnight, S. W. Kantor, and H. Zhu, Annual Tech. Conf., Society of Plastic Engineers, 54, Vol 2, 1594 (1996).
50. D. J. Walsh. J.S. Higgins, and S. Rostini, *Macrocmolecules*, **16**, 388 (1983).
51. F. J. Lu, E. Benedetti, and S.L. Lu, *Macromolecules*, **16**, 1525 (1983).
52. T.K. Ahn and S. Choe, *Polymer (Korea),* **19** (5), 633 (1995).
53. E.C. Chenevey, J. Edward, and B.C. Ward, Celanese Corporation internal communications.
54. D.J. Williams, L. Leung, F.E. Karasz, W.J MacKnight, and M. Jaffe, US Patent 4,973,629 (1990).
55. P.N. Chen, M. Glick, R.H. Vora, and M. Jaffe, US Patent 5,137,985 (1992).
56. P. Musto, L. Wu, F.E. Karasz, and W.J. MacKnight, *Polymer*, **32**, 3 (1991).
57. Y. Lu, J. Chen, H. Cui, and H. Zhou, *J. Chil. Chem. Soc.*, **54** (4), 331 (2009).
58. B. Gordon, R.J. Kumpf, P.C. Painter, *J. Polym. Sci., Polym. Chem*, **26** (7), 1689 (1988).
59. W.J. Harris and W. F. Hwang, US Patent 5,098,985 (1992).
60. Y.P. Chen, D.H. Chen, C.A. Arnold, D.A. Lewis, J.F. Pollard, J.D. Graybeal, T.C. Ward, and J.E. McGrath, Am. Chem. Soc., *Div. Polym. Chem. Polym. Prepr.*, **29** (2), 370 (1988).
61. S. Makhija, E. M. Pierce, T.K. Kwei, and F. Liu, *Polymer Eng. and Sci.*, **30** (13), 798, (1990).
62. G Calundann, F. Herold, E.C. Chenevey, and T.S. Chung, US patent 5,290,884 (1994).
63. M. Gruender, *High-PV wear study of six high performance wear grade engineering plastics.* White Paper. PBI Performance Products, Inc. (2002)
64. Y. Wang, M. Gruender, T.S. Chung, 'Pervaporation dehydration of ethylene glycol through polybenzimidazole (PBI)-based membranes', *J. Membr. Sci.*, **363**, 149 (2010).

65. Y. Wang, T.S. Chung, B.W. Neo, M. Gruender, 'Processing and engineering of pervaporation dehydration of ethylene glycol via dual-layer polybenzimidazole (PBI)/polyetherimide (PBI) membranes', *J. Membr. Sci.,* **378**, 339–350 (2011).
66. K.Y. Wang, T.S. Chung, R. Rajagopalan, 'Dehydration of tetrafluoropropanol (TFP) by pervaporation via novel PBI/BTDA-TDI/MDI co-polyimide (P84) dual-layer hollow fiber membranes', *J. Membr. Sci.*, **287**, 60 (2007).
67. Y. Wang, M. Gruender, T.S. Chung, 'Polybenzimidazole-based membranes for the dehydration of organic liquids via pervaporation', United States Patent Application Publication Pub. No. US2011/0266222 A1, Pub. Date Nov 3, 2011.
68. Y. Wang, T.S. Chung, M. Gruender, 'Sulfonated polybenzimidazole membranes for pervaporation dehydration of acetic acid', *J. Membr Sci.*, **415–416**, 486–495 (2012).
69. K. Y. Wang, T. S. Chung, 'Fabrication of polybenzimidazole (PBI) nanofiltration hollow fiber membranes for removal of chromate', *J. Membr. Sci.*, **281**, 307–315 (2006).
70. Kai Yu Wang, Tai-Shung Chung, Jian-Jun Qin, 'Polybenzimidazole (PBI) nanofiltration hollow fiber membranes applied in forward osmosis process', *J. Membr Sci.*, **300** (1–2), 6–12 (2007).
71. R. P. Singh, X. Li, K. Dudeck, B.C. Benicewicz, K. A. Berchtold, 'Micro-structural optimization of polybenzimidazole-based membranes for H2/CO2 separation at elevated temperatures', North American Membrane Society meeting (2012).
72. G. Krishnan, I. S. Jayaweera, 'Fabrication and scale-up of polybenzimidazole (PBI) membrane based system for precombustion based capture of carbon dioxide', 2011 NETL CO2 Capture Technology Meeting, August 2011.
73. K. Iura, Y. Onishi, T. Itoh, US patent 5549946 (1996).
74. Y. Onishi, K. Iura, A. Maeno, T. Itoh, K. Tsuji, US patent 5674614 (1997).
75. J. Moore, G. Copeland, M. Gruender, 'Polybenzimidazole/polyacrylate mixtures', US patent Application Serial No. 12/957601.
76. J. Moore, G. Copeland, M. Gruender, US patent 8475925 (2013).

Index

ab initio methods, 52
accelerated solvent extraction (ASE), 33
acoustic microscopy (AM), 85–6
aerospace applications, 99–104
 characterisation of HTPBs, 104–9
  atomic oxygen erosion yield, 105
  recommended tests procedures for replacement aliphatic PU coatings, 109
 HTPBs in IKAROS (Interplanetary Kite-craft Accelerated by Radiation of the Sun) spacecraft, 102
 HTPBs in NASA polymeric spherical electrostatic shield, 102–4
  charged spacecraft and six positively and negatively charged spheres, 102
  room temperature DC dielectric and ultimate tensile strengths, 103
  structure of TOR-NC, 103
 HTPBs in Solar Impulse aircraft, 99, 101
 summary of SAE polymers and blends standards, 100–1
American Membrane Technology Association (AMTA), 73
Arrhenius equation, 85
atomic force microscopy (AFM), 28, 81, 85–6
atomic oxygen (AO), 102–3
atomistic calculations, 28
attenuated total reflectance (ATR), 79, 98
Auger spectroscopy, 25
automatic continuous online monitoring of polymerisation (ACOMP), 33

bi-layer cells, 88
bismaleimides, 9
blending miscible high temperature polymers, 132
blending polymers
 challenges, 45–9
  examples of polymer-1 degradative effects on polymer-2 in HTPB, 47–8
bulk heterojunction (BHJ) cells, 88

calorimetry, 159
carpet shape, 104
Celazole, 183, 200–2
Certified Reference Materials (CRM), 26
characterisation methods
 high temperature polymer blends, 14–52
  challenges in blending polymers, 45–9
  chemical, thermal, mechanical and radiation-induced degradation, 33–42
  chemical constitutions and molecular weights, 29–3
  future trends, 50–2
  methods of polymer characterisation, 23–6
  polymer blends characterisation, 26–8
  stabilisation, 42–5
  summary, 49–50

chemical degradation, 34
chromatography, 24, 31
cloud-point curve (CPC), 27
Cole-Cole plot, 84–5
commercial liquid crystal polymer (LCP) blends, 157–60
  commercial products, 158–9
  patents, 157–8
commercial products, 158–9
compatibilisation, 27
Compression Test, 25
cyclic voltammetry (CV), 97

de-polymerisation, 34–5
differential scanning calorimetry (DSC), 35, 80, 167
differential thermal analysis (DTA), 35, 85
differential thermogravimetry (DTG), 34
diffractometry, 24
diphenylisophthalate (DPIP), 178
direct injection moulding, 169
donor-acceptor-donor (DAD), 98
dried blend resin, 188
dye-sensitised solar cell (DSSC), 92
dynamic mean field density functional theory, 52
dynamic mechanical analysis (DMA), 135
dynamic mechanical measurements, 152
Dyneema, 16–17

E-tek, 84–5
Ektar, 41
electrochemical impedance spectroscopy (EIS), 84
electrochemical voltage spectroscopy (EVS), 97
Electron Backscattered Diffraction Analysis, 26
Electron Probe Microanalysis, 26
Electron Spectroscopy for Chemical Analysis (ESCA), 98
electron spin resonance (ESR), 42–3
energy dispersive X-ray spectrometers, 26
Environmental Stress Cracking Resistance (ESCR), 27–8
equivalent sun days (ESD), 107
evaporation of solvent through surface of encapsulating and induced alignment (ESSENCIAL), 89
exhaustiveness, 142–3
Extem, 22
external quantum efficiency (EQE), 95, 96–7

Fenton's test, 85
field emission microscopes (FEM), 73–4
fill factor, 95
Flexural test, 25
Flory-Huggins formalism, 3
Fourier transform-infrared (FT-IR), 167
  analysis, 24
  spectroscopy, 79
Fox equation, 187
fuel cell (FC) membrane applications, 74–9
  Nafion structure, 75
  polymer blends used for PEM, 76–8

gas chromatography-mass spectrometry (GC-MS), 35
gas permeability, 82
gel permeation chromatography (GPC), 31, 99

heat deflection temperature (HDT), 185
high modulus aramides (HMA), 192
high resolution nuclear magnetic resonance techniques (HR-NMR), 28
high temperature polymer blends, 1–10, 18–24
  amorphous siloxane-polyetherimide copolymer SILTEM, 23
  characterisation for specific applications, 70–110
    aerospace applications, 99–104, 104–9
    fuel cell (FC) membrane applications, 74–9
    HTPBs arranged in order of increasing continuous use temperature, 71–2
    membrane applications, 73–4

polymer solar cells (PSCs), 93–9
solar cell (SC) applications, 86–93
characterisation methods, 14–52
challenges in blending polymers, 45–9
chemical, thermal, mechanical and radiation-induced degradation, 33–4
example of rigid-rod and kink-forming monomeric structures, 16
future trends, 50–2
HTPBs arranged in order of increasing continuous use temperature, 19–20
methods of polymer characterisation, 24–7
PA blends with specialty polymers, 17–18
polymer blends characterisation, 27–9
summary, 49–50
world plastics production, 15
chemical constitutions and molecular weights, 31–3
characterisation, 30
FC applications, 79–86
four categories of amorphous and semi-crystalline polymers, 18
general principles of polymer blending, 2–3
immiscible blends, 5–9
liquid crystal polymers (LCPs) as reinforcement, 141–60
blends of isotropic and anisotropic thermotropic polymers, 150–2
commercial LCP blends, 157–60
future trends, 160
liquid crystals, 144–5
polymer liquid crystals, 145–50
processability of LCP and thermoplastic blends, 152–4
researching, 142–4
structure–property relationship of LCP blended materials, 155–7
polysulfones as reinforcement, 165–72
issues in blending polysulfone with other HTP, 166–8
physical properties, 168–9
structure and properties, 165–6
thermoset mixtures, 169–71
stabilisation, 42–5
benzofuran-2-one, 45
example of labile hydrogens, 44
example of plastics stabilisers, 44
thermodynamics, 3–5, 130–9
blending miscible high temperature polymers, 132
chemical structure of poly(2,2′(m-phenylene)-5-5′ bibenzimidazole), 5
liquid crystal polymer blends, 135–7
molecular composites, 137–8
PBI blends, 133–5
polyimide blends, 135
highest occupied molecular orbital (HOMO), 94
hydrolysis, 34–5
hydroquinone (HQ), 136
6-hydroxynapthoic acid (HNA), 135–6
HyperBlocks, 52
HyperMacs, 52

IKAROS (Interplanetary Kite-craft Accelerated by Radiation of the Sun) spacecraft, 102
immiscible blends, 5–9
individual polymer components, 79–81
infrared (IR) spectra, 192
infrared reflection-absorption spectroscopy (IRRAS), 107
infrared studies, 133–4
injection-moulded articles, 155
Innegra, 16–17
internal quantum efficiency (IQE), 96–7
International Atomic Energy Agency, 41–2
International Space Station (ISS), 102–3
interpenetrating polymer network (IPN), 108
interphase, 27
inverse gas chromatography (IGC), 33

ion-exchange capacity (IEC), 83
Irgafos, 44–5
irradiation, 40
ISO Standards, 24–26

Japan Aerospace Agency (JAXA), 102

Kapton, 106
Kevlar, 16–17, 137, 146–7

Lexan SLX, 22
light scattering, 29, 31
linear Rule of Mixtures, 187
liquid crystal device (LCD), 102
liquid crystal polymer blends, 135–7
liquid crystal polymer (LCP), 7–9
  reinforcement in high temperature polymer blends, 141–60
    blends of isotropic and anisotropic thermotropic polymers, 150–2
    commercial LCP blends, 157–60
    future trends, 160
    liquid crystals, 144–5
    polymer liquid crystals, 145–50
    processability of LCP and thermoplastic blends, 152–4
    researching, 142–4
    structure–property relationship of LCP blended materials, 155–7
liquid crystals, 144–5
  nematic structure of p-azoxyanisole, 145
  transition temperatures of p-azoxyanisole, 146
liquid separation, 208
low-angle laser light scattering (LALLS), 32
low Earth orbit (LEO), 42, 104
low molecular weight liquid crystals (LMWLC), 137
lowest unoccupied molecular orbital (LUMO), 94
lyotropic polymer solutions, 146–7

magic angle spinning (MAS), 79–80
mass spectrometry, 31
mass spectroscopy, 31
Matrimid, 179–80
maximum power point, 95
mechanical degradation, 37–40
melt rheology, 81
melt viscosity, 7–8
membrane water uptake, 83
methanol (MeOH) permeability, 82–3
methylene chloride, 193
metrology, 23
microanalysis, 25
micrography, 25
microscopy, 25
microwave-assisted extraction (MAE), 33
miscible polymer blend, 3, 27
molecular composites, 6, 137–8
molecular dynamics (MD), 52
Monte Carlo (MC), 52
multi-junction cells, 90
multi-scale molecular modelling (MSMM), 52
multi-wall carbon nanotubes (MWCNT), 156–7
multiphase systems, 50

Nafion, 51, 75, 110
nanofiltration, 73
naphthalate dicarboxylate (NDC), 155
NASA polymeric spherical electrostatic shield, 102–4
natural gas direct carbon fuel cell (NGDC-FC), 74
near-infrared (NIR) spectroscopy, 33, 79
nematic-to-isotropic transition, 144–5
nuclear magnetic resonance (NMR), 31, 79–80, 98
Nylon 12, 158
Nylon 6 system, 6

open circuit voltage, 95
optical microscopy (OM), 81, 85–6
organic solar cells (OSC), 86–7

p-hydroxybenzoic acid (HBA), 135–6
partial discharge erosion (PDE), 106
patents, 157–8
phase separation, 169–70, 171
photovoltaic (PV) devices, 86
poly (2,2′ (m-phenylene)-5-5′ bibenzimidazole) (PBI) blends, 133–5

chemical structure of PBI polymer, 133
chemical structure of Ultrem polyetherimide, 133
phase separation diagram for PBI/Ultrem blends, 134
polyamide-imides (PAI), 196
polyarylates (PA), 192
polybenzimidazole (PBI), 166
high temperature applications, 206–8
composites and coatings, 208
examples of other applications, 207–8
fabrics made of fibre blends, 207
moulded products, 206–7
high temperature polymers and blends, 174–209
future trends, 208–9
PBI–poly (bisphenol-A carbonate) (PC) and PBI–polybenzoxazole (PBO) blends, 197–8
PBI–poly(4-vinyl pyridine) (PVPy) and other blends, 198–9
PBI–polyarylate (PA) blends, 192–4
PBI–polyaryletherketone (PAEK)–PEI blends, 190–2
PBI–polyetheretherketone (PEEK) blends, 202–5
PBI–polyetherimide (PEI) blends, 188–90
PBI–polyetherketoneketone (PEKK) blends, 183–8
PBI–polyimide (PI) and PBI–polyamide-imide (PAI) blends, 195–7
PBI–polysulfone (PS) blends, 194–5
processing of PBI, 177–9
polymerisation structure, 179
structure of PBI polymer, 175
TGA curve of 100 mesh PBI Air, 176
TGA curve of 100 mesh PBI N2, 176
polybenzimidazole (PBI) blends, 179–83
polyaryletherketone structures, 182
polybenzimidazole (PBI) commercial products, 199–206
commercial PBI fibres, 199–200
properties of PBI staple fibres, 200
compression moulded products, 200–2
Celazole U-60 typical properties, 201–2
solutions, 205–6
PBI Gold, 199–200
PBI Matrix, 199–200
PBI–poly (4-vinyl pyridine) (PVPy) blends, 198
PBI–poly (bisphenol-A carbonate) (PC) blends, 197
PBI–polyamide-imide (PAI) blends, 196–7
PBI–polyarylate (PA) blends, 192–4
PBI–polyaryletherketone (PAEK)–PEI blends, 190–2
PBI–polybenzoxazole (PBO) blends, 198
PBI–polyetherimide (PEI) blends, 188–90
T_g vs % PBI, 190
PBI–polyetherketoneketone (PEKK) blends, 183–8
PBI–polyimide (PI) blends, 195–6
PBI–polysulfone (PS) blends, 194–5
PBI powder resin, 186
polyetherimide, 133
polyethylene naphthalate (PEN), 155
polyimide blends, 135
polyimides (PI), 166
polymer alloys, 27
polymer blending, 2–3
polymer liquid crystals, 145–50
lyotropic polymer solutions, 146–7
normal stress behaviour for PBLG in m-cresol, 148
viscosity at various shear rates vs concentration and chemical structure of PBLG, 148
thermotropic liquid crystal polymers, 147–50
Polymer Reference Interaction Site Model (PRISM), 29
polymeric solar cells (PSC), 87
polymers, 24

poly(phthalazinoneethersulfoneketone) (PPESK), 22
polysulfones
  reinforcement in high temperature polymer blends, 165–72
    issues in blending polysulfone with other HTPs, 166–8
    physical properties, 168–9
  structure and properties, 165–6
  thermoset mixtures, 169–71
    chemical structure of bisphenol A dicyanate, 170
positron annihilation lifetime spectroscopy (PALS), 35, 73–4
power, 95
power conversion efficiency (PCE), 95
protective clothing, 209
proton conductivity, 84–5
pulse-induced critical scattering (PICS), 27
pyrolysis-gas chromatography-mass spectrometry (GC-MS), 80–1
pyrolysis gas chromatography (PGC), 35

quantum dots (QD), 90–1

radiation degradation, 40–2
Raman spectroscopy, 79, 167
reference electrode (RE), 97
relative humidity, 83
rheology, 148–9, 159
rigid backbone polymers, 146–7

scanning electron microscopy (SEM), 8, 26, 73–4, 81, 85–6
secondary ion mass spectroscopy, 25
selected-area electron diffraction (SAED), 97
self-reinforcing composite, 143
self-reinforcing polyphenylenes (SRP), 17
semi-interpenetrating networks, 9
short circuit voltage, 95
SILTEM, 23
single angle neutron scattering (SANS), 108
single layer cells, 88
single-walled carbon nanotubes (SWCNT), 92–3, 156–7
size exclusion chromatography (SEC), 80, 99
small angle neutron scattering (SANS), 29
small angle X-ray scattering (SAXS), 98–9, 189
solar cell (SC) applications, 86–93
  efficiency of diverse types of inorganic or organic solar cells, 87
    active bi-layer of $PC_{60}BM$, 89
    current–voltage dependence, 94
    donor polymers structure, 90
    maximum power point, 95
    n-type ZnO nanocones penetrating p-type PX CdTe matrix on transparent ZnO, 92
    PEDOT:PPS multi-layered structure, 88
Solar Impluse aircraft, 99, 101
solid-state nuclear magnetic resonance, 189
solid-state polymerisation (SSP) process, 178
solution viscosity theory, 32
sonification-assisted extraction, 33
spectroscopy, 25
spinodal decomposition (SD), 108
spiral flow, 153
  pure Ultrem vs HIQ/Ultrem blends, 154
standard test conditions (STC), 96
sterilisation, 41
structure–property relationship
  LCP blended materials, 155–7
    example of good adhesion in LCP blend, 156
    SEM micrograph of LCP fibrils both core and skin regions, 156
sulfonation, 195
supercritical fluid extraction (SFE), 33
  Superex Polymer, 158–9

tandem cells, 90
temperature rising elution fractionation (TREF), 32–3

terephthalic acid (TA), 136
tetraaminobiphenyl (TAB), 178
tetrahydrofuran, 193–4
theoretical model, 3
thermal degradation, 34–8
  molecular structure of poly(ether-ether-ketone) and Vectra LCP, 36
  poly(2,6-dimethyl-1,4-phenylene oxide, 37
  polycarbonate of bis-phenol-A, 35
  pyromellitic di-anhydride (PMDA), 38
thermal mechanical analysis (TMA), 192
thermal volatilisation analysis (TVA), 35
thermodynamics, 3–5
  high temperature polymer blends, 130–9
    blending miscible high temperature polymers, 132
    liquid crystal polymer blends, 135–7
    molecular composites, 137–8
    poly (2,2′ (m-phenylene)-5–5′ bibenzimidazole) (PBI) blends, 133–5
    polyimide blends, 135
thermogravimetric analysis (TGA), 25, 80, 99
thermomechanical degradation, 37–40
thermoplastic polyimide (TPI), 21
thermoset mixtures, 169–71
thermotropic liquid crystal polymers, 147–50
thermotropic polymers
  blends of isotropic and anisotropic, 150–2
  mechanical properties of LCP and polymer blends, 151–2
  miscibility and compatibility, 151
transmission electron microscopy (TEM), 81, 85–6, 192
twin-screw extruder (TSE), 34
two-stage melt condensation, 178

ultrafiltration, 73
ultraviolet (UV) irradiation, 42
ultraviolet (UV)-visible, 42–3, 96–7, 98
Ultrem, 22–3, 133, 188–9
US National Renewable Energy Laboratory (NREL), 90

vacuum ultraviolet (VUV), 106
Vectra, 36
Vectra A950 thermotropic polyester, 159–60
Victrex, 22–3
viscometry, 31
viscosity
  behaviour, 147
  ratio, 152–3
Vogel-Fulcher-Tamman-Hesse (VFTH), 85

wavelength dispersive spectroscopy, 26
Wayback Machine, 143
Web of Science, 142–3
weld lines, 153–4

X-ray diffraction (XRD), 41, 98, 159, 193
X-ray photoelectron spectroscopy (XPS), 25, 41, 98

Young's modulus, 152, 168, 169

Kolibri 360 GmbH
Ferdinand-Jühlke-Straße 7
99095 Erfurt, Deutschland
produktsicherheit@kolibri360.de